From a 1975 oil-on-canvas Ethiopian painting by local artist Ato Tesfaye Tave. In what was locally called a Queen of Sheba style, it portrays activities and adventures of the smallpox program staff. Pictures include those of smallpox patients, vaccination using either a bifurcated needle or a jet injector, and searching for possible cases. At the center top is the SEP emblem of the Smallpox Eradication Program. The original is 54 x 29 inches (137 x 74 cm). It is in the author's collection.

CRITICAL ACCLAIM FOR
SMALLPOX—THE DEATH OF A DISEASE

"Like the Pied Piper, D. A. Henderson came recruiting to the London School of Hygiene and Tropical Medicine. For those of us who followed, it was the beginning of an experience that shaped a lifetime—of lasting friendships and public health adventure so well described in this, the Piper's book—a monumental work by an exceptional leader."
> —David L. Heymann, MD, assistant director-general for
> public health security and polio eradication,
> World Health Organization

"Hundreds of millions contracted and died from smallpox and hundreds of thousands of smallpox eradicators were in the 'army' that conquered this awful disease. But only D. A. Henderson could write a remarkable book like this from the perspective of the general who led that historic campaign. My hope is this story inspires those today who are fighting to add more diseases to the dustbin of history: polio, malaria, and Guinea Worm."
> —Larry Brilliant, MD, MPH, vice president and
> chief philanthropic evangelist, Google

"In *Smallpox*, Dr. Henderson skillfully takes the reader from the earliest days of this plague through the comprehensive and complex global program that ultimately led to its successful eradication. He then offers a sobering yet intriguing account of relevant threats since then. Although he is generous in sharing the credit for this heroic accomplishment with a huge number of collaborators, his guiding presence is evident throughout this fascinating and wonderfully accessible account."
> —Andrew A. Sorensen, distinguished president emeritus,
> University of South Carolina

"The eradication of smallpox was the greatest public health accomplishment of the twentieth century. But the lessons of the story...extend far beyond global health. D. A. Henderson's *Smallpox—The Death of a Disease* makes me optimistic about our future."

    —Elizabeth Fenn, author of
        Pox Americana: The Great Smallpox Epidemic of 1775–82
        and associate professor of history, Duke University

"*Smallpox* describes how an international team of idealistic but pragmatic doctors and public health practitioners overcame...complex technical, logistical, and political challenges to eradicate an epidemic disease that had tormented humanity for millennia. It is a gripping and inspiring tale, punctuated by repeated setbacks and crises, and told with frank immediacy by the American physician who led the ultimately successful global campaign."

    —Jonathan B. Tucker, PhD, author of
        Scourge: The Once and Future Threat of Smallpox

"From his unique perspective as the leader of the eradication team, D. A. Henderson tells the dramatic story of smallpox, the disease that killed more people than any other in history. Its eradication is, perhaps, man's most magnificent accomplishment. Dr. Henderson's account is spellbinding, and the lessons are the foundation of worldwide disease control."

    —Parker A. Small Jr., MD, professor emeritus,
        University of Florida College of Medicine, and
        charter member, National Vaccine Advisory Committee

SMALLPOX

D. A. Henderson, MD

SMALLPOX
THE DEATH OF A DISEASE

The Inside Story of
Eradicating a Worldwide Killer

FOREWORD BY RICHARD PRESTON

Prometheus Books

59 John Glenn Drive
Amherst, New York 14228–2119

Published 2009 by Prometheus Books

Inquiries should be addressed to
Prometheus Books
59 John Glenn Drive
Amherst, New York 14228–2119
VOICE: 716–691–0133, ext. 210
FAX: 716–691–0137
WWW.PROMETHEUSBOOKS.COM

13 12 11 10 09   5 4 3 2

Library of Congress Cataloging-in-Publication Data

Henderson, Donald Ainslie, 1928–
    Smallpox : the death of a disease : the inside story of eradicating a worldwide killer / by D.A. Henderson ; foreword by Richard Preston.
        p. cm.
    Includes bibliographical references and index.
    ISBN 978–1–59102–722–5 (cloth : alk. paper)
    1. Henderson, Donald Ainslie, 1928– 2. Smallpox—History—20th Century.
I. Title.
    [DNLM: 1. Smallpox—epidemiology. 2. History, 20th Century. 3. Smallpox—history. 4. Smallpox—prevention & control. WC 585 H496s 2009]

RA644.S6H46 2009
362.196'9120092—dc22

                                                                    2009010087

Printed in the United States of America on acid-free paper

To my family, who provided both unflagging support and tolerant acceptance of the frequent absences of a father during the many challenging years of the smallpox saga: Nana, Leigh, David, and Douglas

CONTENTS

CHAPTER 11. LESSONS AND LEGACIES OF SMALLPOX ERADICATION   301

FOREWORD

by Richard Preston

On May 8, 1980, the World Health Organization (WHO) declared smallpox eradicated. No cases of smallpox had occurred anywhere on earth for more than two years; the smallpox virus, the cause of the worst infectious disease in history, had been removed from the human species. As much as anyone, D. A. Henderson was responsible for the eradication of smallpox. (The initials stand for Donald Ainslie; everyone, including his wife, Nana, calls him D. A.) He led the WHO's Smallpox Eradication Unit from its inception in 1967 through 1977 when the last naturally occurring human case of the disease occurred.

Many people, of course, contributed to this triumph. Those who worked in the project called it simply the Smallpox Program. I will call it the Eradication. It was one of the noblest and best things that we have ever done, as a species. The Eradication ultimately involved hundreds of thousands of people. They worked in many countries and they came from all walks of life, from top officials at the WHO's Geneva headquarters to village health workers who could not read or write but who could diagnose a case of smallpox faster than most physicians in the United States or Europe. When the Eradication began, in 1967, the World Health Assembly set a goal of ten years for the elimination of smallpox. The Eradicators referred to this goal as Target Zero. It meant no case of smallpox anywhere on earth. They missed the goal by only nine months.

As a virus, smallpox is an exceedingly small, infectious biological particle, a parasite capable of making copies of itself inside the cells of a host organism. It undergoes self-replication when it gets inside cells of its natural host, *Homo sapiens*. The human species is the only natural host of smallpox; no other organism harbors the virus. If a smallpox particle makes its way inside a human cell, it takes over the cell's machinery and turns the cell into a factory for making more smallpox particles.

In the last hundred years of its existence, smallpox is thought to have killed at least half a billion people. All the wars on the planet during that time killed perhaps 150 million. In the contest of Smallpox vs. War, War lost. Smallpox killed roughly one-third of the unimmunized people it infected, and the disease was grisly.

Once a person was infected with smallpox, there was an incubation period of around ten days before the person became noticeably sick. Then the person got a high fever and severe aching pains. After two to three days, the patient would begin to develop a rash. The rash appeared on the face, hands, and feet, and quickly rose into pustules. Smallpox pustules were hard, pressurized blisters filled with a clear, faintly opalescent pus. The pain of the smallpox pustular rash was virtually unbearable. If the pustules merged into sheets, which was called a confluent rash, the patient was very likely to die. They died of shock.

Doctors could diagnose a case of it with their eyes closed, by sense of touch, simply by running their fingertips over the pustules: they were rock-hard and were supposed to feel like buckshot embedded in the skin. Smallpox had a unique odor. A smallpox patient emitted a sickly-sweet stench that seemed to come straight out of the skin and would fill a ward.

As many as 5 percent of smallpox cases were manifested in a form called hemorrhagic smallpox. This form of smallpox disease was invariably fatal and was a bloody mess. Bleeding occurred under the skin, which didn't develop pustules but had a darkened, corrugated appearance. The skin of a hemorrhagic smallpox patient could look as if it had suffered third-degree burns and could slough off in sheets when nurses tried to move the patient. These patients had hemorrhages from the mouth and nose, intestinal tract, and urinary tract. The whites of the eyes of a hemorrhagic smallpox victim would turn bright red from blood leaking into the eyes. Smallpox was a monster.

The task of reaching Target Zero required a number of technical and organizational achievements. First, there had to be a vaccine—cheap, stable, and effective. This sine qua non first step toward the Eradication was taken in 1796 when the English country doctor Edward Jenner scratched the arm of a boy named James Phipps with some pus containing cowpox virus, a virus closely related to smallpox. Several months after he'd "vaccinated" the boy, he scratched the boy's arm with pus that contained smallpox virus. The boy didn't get sick. Jenner had invented the first vaccine. Today's smallpox vaccine is similar to the one Jenner used. It is a live virus—not smallpox—called *vaccinia*. When a small amount of *vaccinia* virus is introduced into a person's skin—by scratching the arm with it, just as Jenner did—the person develops a blister on the arm and becomes strongly immune to a smallpox virus infection.

One hurdle the Eradicators faced was skepticism within the scientific community about the feasibility and practicality of eradicating an infectious disease. To some biologists, the goal sounded quixotic, even wrongheaded. Earlier attempts to eradicate microscopic pathogens had failed, most notably an attempt to eradicate malaria. The evolutionary biologist and essayist René Dubos, writing in 1965, just two years before the start of the Eradication, expressed it this way: "Even if genuine eradication of a pathogen or virus on a worldwide scale were theoretically or practically possible, the enormous effort reaching the goal would probably make the attempt economically and humanly unwise."

René Dubos was being reasonable, and he was dead wrong. As it turned out, the cost of each smallpox vaccination given by the Eradication came to around ten cents. That includes all costs, including manufacturing, transportation, and staff salaries. By contrast, the actual *cost* of smallpox disease to the world was hard to calculate, given the fact that about 2 million people were dying of it every year, but certainly the cost amounted to more than a billion dollars yearly.

A major technical step toward the Eradication was the invention, by Dr. Benjamin Rubin, in 1965, of the bifurcated needle. It was a piece of sharpened steel wire that had a double point on one end, looking like a tiny olive fork; it held a very small droplet of the vaccine between its points.

You could vaccinate someone by making fifteen light punctures on the person's skin. It was efficient, using very little vaccine to get the job done. The needle got a hundred vaccinations out of a vial of vaccine, whereas previously a vial had yielded only twenty-five. And the bifurcated needle was cheap. Each needle cost a quarter of a cent. ("We boiled them and reused them many times," Henderson says.)

At first, governments wanted to do mass vaccination, vaccinating every person in a country or region affected by smallpox. Mass vaccination alone was slow and often unsuccessful. The Eradicators invented the technique of ring vaccination, which proved to be powerfully effective. Wherever an outbreak was found, the Eradicators swooped in and vaccinated every person they could find in a community or ring of houses around the outbreak. In this way, they surrounded each outbreak of smallpox with a wall of humans who were immune to the virus. The virus would burn itself out inside the ring and would vanish in that spot. To succeed, ring vaccination required relentless, meticulous surveillance for new cases.

Finally, and not least importantly, the Eradication required organizational creativity. This is where D. A. Henderson's genius for management is apparent. Henderson created an incredibly flat management structure. Normally only nine people, including secretaries, staffed the headquarters offices in Geneva. They managed a program bigger than some Fortune 500 companies, with up to one hundred and fifty thousand employees and staff at a given time who operated around the world, sometimes in countries that were in the middle of civil war. Henderson and his team bypassed many of the WHO bureaucrats. Henderson created a semi-invisible parallel organization in the WHO, working outside the bureaucracy's unwieldy regional layers. He staffed it with bright, self-motivated people whom he trusted. He delegated relentlessly. He had to: there was no telephone contact, no fax, not even Telex contact with Eradicators working in the field. The delegation of tasks and authority went all the way down: Henderson and his colleagues assigned important surveillance work to village health workers who were not always able to read and write but were given authority and responsibility to search for and report cases of smallpox.

The Eradicators eventually squeezed *Variola major* into a final stand on Bhola Island, in the Bay of Bengal, in Bangladesh. It made its last natural appearance in a little girl named Rahima Banu, who survived. At last report, she was still living on Bhola Island and was married with children. In 1976, the Eradicators found the last case of *Variola minor* in Somalia, in a cook named Ali Maow Maalin. Smallpox had been eradicated. Or had it?

At the outset of the Eradication, there were two different definitions of eradication. One was the notion that eradication meant the removal of smallpox virus from the human species; the complete severance of the parasite from its host. The second definition of eradication meant the complete destruction of the virus; the removal of smallpox from the universe; the total extinction of smallpox as a life-form.

Though smallpox had been taken out of the human body, it remained alive in captivity. Smallpox can be frozen, held in suspended animation in a vial. A deeply frozen lump of smallpox the size of a pencil eraser is thought to remain viable and infective for centuries or longer. After the Eradication, the World Health Organization called for the destruction of all existing laboratory stocks of smallpox virus. At least seventy-five laboratories around the world held frozen samples of scabs and pus that contained live smallpox. Most laboratories readily complied.

The WHO sanctioned just two national laboratories to hold smallpox frozen for an indefinite period of time, pending the destruction of even these stocks of the virus. One lab was the Centers for Disease Control, a US federal agency in Atlanta, Georgia; the other was the Research Institute for Viral Preparations in Moscow.

Meanwhile, however, the US and British national intelligence services developed evidence that the former Soviet Union had been developing smallpox as a biological weapon. As the Soviet Union collapsed in 1990 and 1991, a series of defectors emerged. They had been working in a secret Soviet program known as Biopreparat. Biopreparat, as it turned out, had been developing biological weapons in research and weapons-production facilities scattered across the length and breadth of the Soviet Union. The biowarfare program of the Soviet Union was large and complicated. Smallpox was one of the crown jewels of the program, and a lot of research work was going into smallpox. One of the defectors, Dr. Kanatjan Alibekov, who changed his name to Ken Alibek and who had been a top scientist in Biopreparat, asserted that Soviet bioweaponeers had routinely manufactured and stored liquid weapons-grade smallpox in twenty-ton tanks. According to Alibek, they had grown the smallpox in cell-culture bioreactors, which are special vessels that can be used for making large quantities of a live virus. (A teaspoon of liquid smallpox preparation could, in theory, infect every person on the planet.)

After the Soviet Union fell apart, the Biopreparat laboratories lost much of their government funding, and many bioweaponeers lost their work and had no income. Many former bioweaponeers left the Soviet Union for places unknown. Any of these scientists could have taken a freeze-dried sample of smallpox along in a small plastic vial about the size of a pencil stub. Representatives of the Russian government eventually told American counterparts that not even the Russian government knew the whereabouts of all of the biologists who had left Biopreparat. Because there was virtually no smallpox vaccine left on earth, and because, as the years had passed, fewer people had immunity to smallpox (the vaccination wears off, and childhood smallpox vaccinations had ceased), smallpox had become a de facto weapon of mass destruction.

In 1997, I was invited to speak at the annual meeting of the Infectious Diseases Society of America (IDSA), in San Francisco, along with D. A. Henderson. We had become friends. Henderson had learned of the Soviet smallpox biowarfare efforts. He was especially concerned about the virtual absence of scientists who had apparently done weapons-related research on smallpox. Henderson had been pushing for a public, official destruction of the known stocks of smallpox. He also wanted there to be a public discussion of the danger of smallpox: he wanted the medical community to be aware that smallpox still existed and could be used as a terrorist weapon. I had been writing about biological weapons and, like Henderson, I had come to the conclusion that smallpox was one of the most dangerous potential weapons, of any kind, in the modern world.

The audience at the IDSA consisted of around three thousand infectious-disease specialists, most of them physicians, many of them practicing in emergency-room settings. Henderson's and my speeches did not go over very well at the time, at least in my view. The audience seemed quiet and skeptical. After all, they had thought that smallpox had been eradicated, and most people generally don't like to hear that a big problem exists when things had seemed fine. Even so, Henderson received many more invitations to talk to doctors and he went all over the United States, speaking about the dangers of smallpox.

Yet it all seemed theoretical. In my own talks about bioweapons, I often encountered the question "If bioweapons are so dangerous and supposedly so easy to make and use, why haven't we seen a terrorist attack with a bioweapon?" It was a good question, and I couldn't really answer it.

Then, in September and October 2001, someone mailed small quantities of powdered anthrax to two US senators and a handful of figures in the news media. Five people subsequently died from inhalational anthrax, most of them postal workers who'd breathed minute quantities of the particles in mail-sorting facilities. It implied that a terrorist attack with smallpox could be worse. Anthrax is not contagious and doesn't spread from person to person. Smallpox is contagious; it spreads in the air. The reason why the Soviets were so interested in smallpox as a weapon was because it would be a self-amplifying weapon once it was introduced into a human population. Plutonium can't make copies of itself. Smallpox can.

Henderson played a key role in the subsequent US government decision to create a national stockpile of smallpox vaccine, so that in the event of a smallpox outbreak, every citizen, at least in theory, would be able to get a smallpox vaccination. At the same time, Henderson has kept pushing to have all of the official frozen stocks of smallpox in the United States and Russia destroyed. Henderson doesn't believe that a public destruction of the known stocks would guarantee the extinction of smallpox, since clandestine stocks of the virus could still exist in places, but he feels that the official destruction of smallpox would set a moral and ethical standard for the international community. Any person, group, or nation that continued to hold smallpox after that point would be committing a crime against humanity. The crime could be prosecuted in international courts of law and in the world's public opinion.

The DNA of smallpox has been decoded. The genetic sequence of the virus is known; the smallpox genome is in the public domain. You can look it up on the Internet. Even if all stocks of the virus were destroyed, we—the human species—would still possess the recipe for making it. It might be possible some day to re-create smallpox in a laboratory, although this would not be easy. It would require, at the very least, fast gene-synthesis machines—machines that could string together thousands of letters of DNA into genes rapidly and accurately.

How real is the threat of smallpox today? The truth is, no one knows. The threat of a bioterror event with smallpox is not zero, nor is it a dead certainty. Smallpox lurks in the shadows, a potential nightmare of the future.

None of this detracts from the shining achievement of the Eradication. As a fundamental fact, variola has been removed from our bodies; the worst infectious disease in history is gone. It has meant that an estimated sixty million lives were saved—productive human lives that would otherwise have been lost to smallpox in the decades since the Eradication in 1977. Just as importantly, the lessons learned in the Eradication have been extended to mass vaccinations around the world for other diseases—measles, hepatitis, rubella, and polio—saving millions more lives. The Eradication was arguably the greatest life-saving achievement in the history of medicine, the attainment of Everest in a medical sense. We can all admire it and we can learn how it was done in this book.

Richard Preston
Princeton, 2008

PREFACE

Smallpox has played a pivotal role in every era of human history. No disease has been so greatly feared or worshipped—no disease has killed so many hundreds of millions of people nor so frequently altered the course of history itself. As I was growing up, however, I knew smallpox only as a name, a disease against which all children had to be vaccinated. That abruptly changed in 1947.

Smallpox suddenly appeared in New York City—two smallpox patients were discovered, but no one knew how or where they had acquired the infection. Their movements were traced, and more smallpox patients were discovered. Emergency vaccination programs began—first for the hospital staff and the patients where the cases were isolated and then for residents of the apartments where they had lived. As more smallpox patients were found, the vaccination program extended to other hospitals and to other parts of the city. Eventually, the source was discovered: a visitor from Mexico who had become ill and died five days after his arrival. During his stay in a hotel, 3,000 people from twenty-eight states had booked rooms. Health staff sought to trace and vaccinate all of them. The city was in turmoil. A decision was finally made to vaccinate the entire urban population. Six million people were vaccinated during a four-week period. This massive effort was the response to an outbreak that consisted of only twelve patients, two of whom died.

Berton Roueché, the respected *New Yorker* medical writer, vividly described the evolving events, the threat, and the terror in an article "The Man from Mexico." He quoted from a doctor's description: "The patient often becomes a dripping, unrecognizable mass of pus by the seventh or eighth day of the eruption. The putrid odor is stifling, the temperature often high (107° has been authoritatively reported), and the patient frequently in a wild state of delirium." For me, the pervasive concern and fear of smallpox was startling and yet I had known nothing of this disease until its unexpected appearance in New York.

Fourteen years later, in 1961, I would be assigned national responsibility for dealing with smallpox, should it be imported into the United States. My position was chief of the surveillance section at the US Communicable Disease Center (CDC). It was a time of high anxiety. Major smallpox epidemics were then erupting across India and Pakistan. Travelers flying by jet aircraft were rapidly increasing in number, and some were infected with the smallpox virus. From 1958 through 1960, the disease had been imported into Europe from Asia six times; eleven more importations occurred in 1961. By the end of 1963, twenty-three importations had resulted in nearly 400 cases. Not surprisingly, we had a number of false alarms in the United States—primarily patients with chicken pox. I assumed that it was only a matter of time before we would have to cope with smallpox.

There were two basic approaches to prevention. One was the well-established defensive strategy—continued compulsory vaccination of children, inspection of travelers' vaccination certificates showing that they had been successfully vaccinated within the previous three years, and rapid investigation of all suspected cases. The longer-term approach was to work with other countries to stop smallpox epidemics at their source—to eradicate the disease. Thus began my career adventure with smallpox and smallpox eradication.

A number of important books have been written about smallpox—its scope and its terrible history—by Donald Hopkins (*The Greatest Killer*), Jonathan Tucker (*Scourge*), Elizabeth Fenn (*Pox Americana*), Richard Preston (*Demon in the Freezer*), Horace Ogden (*CDC and the Smallpox Crusade*), David Koplow (*Smallpox: The Fight to Eradicate a Global Scourge*), among others. And I, with four colleagues, documented the story of smallpox eradication in a detailed archival history—a fourteen-hundred-page account (familiarly known as "The Big Red Book"). That book, *Smallpox and Its Eradication,* was published by the World Health Organization in 1988. It is now out of print but is available on the World Health Organization Web site.

There has been no personal account of the dramatic month-to-month story of the eradication of smallpox—from the early years of plotting program strategy to the celebration of eradication, and, finally, to coping with the threat of its use as a biological weapon. In this book, I have attempted to relate how and why smallpox eradication proceeded as it did and the events, the decisions, and the circumstances concerned. In attempting this, I have had to be selective because the story involves casts of thousands on four continents and over a period of nearly fifty years.

This account was written in part with a grant from the Alfred P. Sloan Foundation, which has been supportive of scientists undertaking to relate their own stories—how they came to be scientists and what influenced them to undertake the work they did. Or, in my case, to answer the question: How did I become a public health epidemiologist and the director of a campaign that enlisted more than a hundred thousand staff in a global cause?

THE SHAPING OF A CAREER

Medical practice was the career my parents anticipated for me, primarily for family reasons. We lived in the middle-class Cleveland suburb of Lakewood, Ohio. My mother had been a nurse; my father was a mechanical engineer but one who never really enjoyed his work. They were Canadian immigrants from long-resident Scotch-Canadian families—decidedly Calvinistic. Among my mother's family, the McMillans, there was a strong sense of clan. My mother's eldest brother was a physician and a prominent member of the Canadian parliament. Understandably, he was highly esteemed within the family—thus my interest in medicine and, later, in public service.

After graduating from Oberlin College and the University of Rochester School of Medicine, I began an internship at the Mary Imogene Bassett Hospital, a Columbia University teaching hospital in Cooperstown, New York. Oberlin had opened a new world for me in questioning my Calvinist and Republican Party roots and introducing me to a racially and internationally diverse student body. It was the setting for the development of a special friendship with an attractive young biology major, Nana Bragg, now my wife of more than fifty years. The Rochester experience confirmed my commitment to medicine and stimulated an interest in cardiology. However, Rochester offered no courses in public health. To me, it was a totally unknown field.

PUBLIC HEALTH

When I did enter the field of public health, it was because of a Selective Service requirement. I had been deferred from the draft since leaving high school in 1946, but in 1955, I was informed that I had to serve two years in a uniformed service. The Korean War was over, but a physician draft was still needed to staff the medical services. With only one year of internship, I could visualize a not-very-exciting two years—probably doing physical examinations on military inductees. Unexpectedly, a representative from the CDC visited our hospital to discuss an additional option—the Epidemic Intelligence Service (EIS), a part of the CDC and the Public Health Service. At that time, the CDC was a small, little-known agency of the federal government located in Atlanta, Georgia. No one at the hospital had even heard of the still-new EIS. It had been established in 1951 as an emergency group whose task was to investigate epidemics wherever they might occur. This had been during the Korean War, when the threat of biological weapons—then, as now—was a real concern.

I had little interest in infectious diseases, but I hoped that the EIS might at least provide a productive educational experience. There were only thirty-five places available, and many competing for them. My only special attribute was that I had written a prize-winning paper on the history of an 1833 cholera epidemic in Rochester. Actually, I have to admit that I was more interested in the financial prize than in cholera. However, I found it fascinating to plot the cases by residence and date and to speculate on how the disease was being spread. It turned out to be my introduction to epidemiology. To the CDC recruiters, it was taken as evidence of a special interest in communicable diseases. I was selected and assigned the job of "assistant chief of the EIS." Two months later, the chief EIS officer left to take another job and, suddenly, I was acting chief—obviously underqualified.

DISEASE DETECTIVES

My boss was Dr. Alexander Langmuir, chief of the CDC's epidemiology branch. Already, at age forty-five, he was a legend in public health. Langmuir was a born teacher, a creative thinker, and a hardworking, demanding, sometimes difficult taskmaster. An associate professor of epidemiology at the Johns Hopkins School of Public Health, he had been recruited to the CDC in 1949 to become the CDC's first epidemiologist. Earlier in his career, he had been a field epidemiologist in the New York State Health Department where he acquired a strong belief in "shoe leather epidemiology." This was shorthand for the activities of an epidemiologist who himself left the office to personally investigate epidemics—collecting data and interviewing patients and officials. The converse type was the office-bound "shiny pants" epidemiologist who relied primarily on submitted reports. I became thoroughly indoctrinated as a shoe leather epidemiologist and championed this attribute ever after.

Langmuir's intention was to provide each of the inductees sound training and challenging experiences in the field. His hope was to persuade them that public health epidemiology was an exciting specialty worthy of a career. He was amazingly successful.

In the early 1950s, Langmuir introduced the concept of disease surveillance as a tool in disease prevention. For smallpox eradication, its application proved to be a vital factor in the program's success. He defined surveillance as a program designed to collect, on a routine basis, relevant data pertaining to a disease (including the numbers of cases and deaths); to analyze and interpret this information; and to distribute it to all responsible for control of the disease in question. The emphasis was on understanding the disease—who became ill, how it spread, what could be done about it and with what success. Collecting and using such information in control programs would seem both intuitive and obvious, but it was effectively a new concept when Langmuir began advocating it. In the smallpox eradication program, we were to demonstrate that the early detection and containment of cases and outbreaks stopped smallpox over large areas even when vaccination coverage was poor. This strategy's application was pivotal to the achievement of eradication.

Langmuir founded the EIS, which he thought of as a "medical fire department for epidemics." Newspapers often described us as "disease detectives." Alex believed that, like firefighters, we must always be prepared—ready to travel on just a few hours' notice to help state and local health officials deal with disease outbreaks. Epidemic assistance calls were to take me far afield—to Argentina, Alabama, the Philippines, New Jersey, Yugoslavia, and to Samoa in the South Pacific.

I was soon captivated by the prospect of public health as a career. Every outbreak was unique—how it had occurred and developed, which groups were infected, what were the best ways to deploy community and federal resources, and what could be done to prevent a recurrence. Diverse strands of information and expertise had to be woven together—clinical and epidemiological information about the disease, laboratory studies, sociological factors, food preferences, and even the weather. The stakes could be high: in any epidemic, the clock was ticking while one figured out the problem and what could be done about it. Guess wrong or take too long to decipher the problem and take action, and hundreds, maybe thousands could become sick or die. There was nothing routine about the job, and it demanded a great deal of rapid learning.

SMALLPOX BEGINS TO DOMINATE MY AGENDA

In 1961, as the new chief of the CDC's surveillance section, I was deeply concerned about the potential for smallpox importations into the United States. Creation of a special smallpox unit seemed like a good idea. We needed a better knowledge of how European countries were detecting and containing outbreaks, and so, for most outbreaks, I sent one of our EIS staff as an observer to learn. We needed to be able to more rapidly vaccinate large numbers of people as had been done in the 1947 New York City outbreak. One potential answer was an army-designed, hydraulic-powered jet injector that could shoot a small volume of vaccine through the skin under high pressure. As many as one thousand people per hour could be vaccinated. We worked with the inventor and the US Army to adapt the gun for smallpox vaccination even in areas without electricity. Finally, we undertook a national study in the United States to determine how frequently vaccine complications occurred and how serious they might be. The studies showed that there were many more serious reactions following smallpox vaccination than vaccinations for other diseases. However, because of the high risk of smallpox importations we knew that routine vaccination had to be continued as the most prudent course for the present.

Our growing expertise in smallpox eventually led, in 1965, to our undertaking a program supported by the US Agency for International Development. It called for smallpox eradication and measles control in twenty countries of West Africa over a five-year period. Seven months after this was launched, and partially due to the impetus it provided, the World Health Assembly decided to mount a major effort to eradicate smallpox worldwide. WHO director-general Marcelino Candau demanded that I be assigned as director of the global effort.

So it was that the career of an aspiring medical practitioner was transformed into that of a public health professional. And a responsibility for devising defensive strategies to respond to a smallpox importation gradually grew into a global campaign to eradicate the disease altogether.

AN EXPATRIATE FAMILY IN GENEVA

My career direction inevitably involved my family as well. For eleven years, we lived in Geneva, Switzerland—my wife, daughter Leigh, and sons David and Douglas. They, too, became absorbed by the program. We frequently entertained at home, regularly inviting field staff who were visiting. They were of many nationalities, serving in many different countries. Stories of their challenges and adventures lasted well into the evening. The month-by-month victories and defeats in the program were followed by the family as closely as one might follow football. With the players at dinner vividly relating their experiences, the program took on a special reality.

A regular visitor was Vice-Minister of Health Dimitri Venediktov from the USSR. Soon after we arrived in Geneva, I met him at the World Health Assembly and invited him home to have a charcoal-broiled steak, something then unknown to him. Knowing the delegates' busy schedules, I was surprised that he accepted. We had a delightful evening—and on subsequent visits, he immediately asked what night we would have the steak dinner so that he could arrange the rest of his schedule. Although these were during the difficult years of the cold war, my professional relationships with Venediktov and the Russians were always cordial, and their contributions to the program were critical—hundreds of millions of doses of high-quality vaccine, young field epidemiologists, and a world-class research and diagnostic laboratory.

At the conclusion of the program, I sought to find a way to recognize a remarkable international staff. National staffs in many countries received special recognition by their own governments, but there was no mechanism for WHO to express its appreciation. I decided to create an award, a special certificate of appreciation for all who had served in the field. This could not be done officially, as it would be seen as precedent setting, and getting approval would be time consuming at best. So I undertook to contract for the printing of a certificate with personal funds. It carried the following citation:

The triumph belongs to an exceptional group of national workers and to a dedicated international staff from countries around the world who have shared privations and problems in pursuit of the common goal

SMALLPOX TARGET ZERO

To: _____

one of the international staff who assisted the World Health Organization in this historic venture—the ORDER OF THE BIFURCATED NEEDLE is given as recognition of participation in the great achievement

Geneva, 1976

The bifurcated needle, mentioned above, was a unique, inexpensive, forked needle, invented in 1966 as the program began (see figure 17). It made vaccination far easier and permitted a twenty-five-dose vial of vaccine to be used to vaccinate one hundred people. Target Zero, as cited above, was the slogan we regularly used as we neared the end of the program. It emphasized that the goal was not millions of vaccinations but zero cases of smallpox. For those receiving these certificates, my daughter had the idea of bending the end of the needle in a circle to signify the "0" target. It could be used as a lapel or scarf pin, anchored with a commercial clasp. She painstakingly crafted 750, and these were duly distributed to all international staff who had served in the program. Many are wearing them even today, and on many résumés appear the words "Member of the Order of the Bifurcated Needle."

FROM GENEVA TO BALTIMORE TO THE WHITE HOUSE

In February 1977, the goal of eradication was in sight. I left Geneva for Baltimore and the deanship of the Johns Hopkins School of Public Health. Responsibility for the concluding phases of the program passed to my eminently capable, longtime deputy, Dr. Isao Arita. Smallpox, however, remained a substantial part of my life as I participated in the concluding meetings of the WHO commission that certified eradication and a committee that had oversight for posteradication activities—such as monkeypox studies and policies for follow-up of suspect cases. Seven years were devoted to the part-time writing of an archival history.

After fourteen years as dean, I decided to leave the world of academic administration for teaching and writing. But in 1990, President George H. W. Bush asked that I serve as associate director for life sciences in the White House Office of Science and Technology Policy. All seemed to be remarkably quiet in the smallpox arena during this time, and I assumed—wrongly—that the smallpox saga had ended. New, more difficult issues arose after I left the White House to become deputy assistant secretary for science in the Office of the Secretary of Health and Human Services.

CAN THE SMALLPOX VIRUS BE DESTROYED?

For more than a decade, interest in the smallpox virus all but vanished. The virus resided in just two laboratories in the world—in Russia and the United States—and for a decade no research was being done that utilized the virus—in fact, there was little research of any kind relating to smallpox.

Many countries and a WHO committee, as well as major scientific organizations, concluded that the remaining stocks of smallpox virus in the two laboratories should be destroyed. This would eliminate the possibility of the virus escaping from either location and would discourage work elsewhere in the event there were hidden stocks of the virus. All agreed on destruction in December 1995. But at the last moment, the United States, joined by Russia and the United Kingdom, argued for keeping the virus. Continuing, increasingly bitter arguments and countless meetings failed to resolve the issue of whether to retain or destroy the virus. The controversy promises to be on the world agenda until well into the next decade.

OR WILL IT RETURN AS A BIOLOGICAL WEAPON?

Concern about the possible use of weaponized biological agents was thought to have been allayed in 1972. All countries had signed a biological weapons convention pledging to destroy the weapons they possessed and not to engage in research activities relevant to offensive biological weapons. All countries were thought to have complied. In the mid-1990s, however, it gradually came to light that the Soviet Union had wantonly flouted the treaty with an extensive, highly secret biological weapons program. At the top of their list of preferred agents was smallpox. How many other countries might have also embarked on such programs was unknown.

At that time there was little interest or expertise in medicine or public health—whether in government or in academia—in dealing with possible biological weapons. However, it became increasingly clear that they posed a threat. Thus, in 1998, I established a university policy center to alert the public and the scientific community of the biological weapons threat and to develop relevant policies. As the only policy center with a strong base of medical and public health expertise, we had more to do than we could handle.

After the World Trade Towers were destroyed in 2001, intelligence intercepts suggested that there might be a second event—probably the release of a biological weapon. Smallpox and anthrax were at the top of the list of probable agents. Soon I was commuting daily to Washington to consult with the secretary of the Department of Health and Human Services and his staff. A new office was created reporting directly to the secretary—the Office of Public Health Emergency Preparedness. I was asked to be its director. Once again I was working seven-day weeks with a principal item on the agenda being smallpox.

Countless meetings ensued—discussing questions of vaccine use, how to respond to an outbreak, the possible need for new vaccines or antiviral drugs, possible international responses should outbreaks occur in other countries, and quarantine practices. These were in addition to the interminable discussions about possible destruction of the virus. Smallpox, as a topic, was back from the dead—generating more in the way of papers, meetings, and directives than it ever had.

CODA

When all this began for me in 1961, I had not the slightest inkling that smallpox would be a disease that would preoccupy me for a lifetime. Whatever the quandaries, I return to the basic fact that, for the first time in history, a disease has been eradicated—the most serious of all the pestilential diseases. It is a tribute to the dedication, creativity, and sacrifice of tens of thousands of health workers from around the world, working collaboratively under the aegis of the World Health Organization. Without them and without this organization, eradication could never have been achieved.

We are only beginning to realize the potential of public health and to explore new horizons in research, understanding, and application. It is a field begging for fresh, resourceful ideas and a new generation of professionals who are not constrained by "knowing" what can't be done. So it was with so many who contributed so much in making smallpox eradication a possibility. Their stories, as well as mine, constitute the heart of this book.

Chapter 1

THE DISEASE, THE VIRUS, AND ITS HISTORY

"Small pox was always present, filling the churchyard with corpses, tormenting with constant fear all who it had not yet stricken, leaving on those whose lives it spared the hideous traces of its power, turning the babe into a changeling at which the mother shuddered, and making the eyes and cheeks of the betrothed maiden objects of horror to the lover."

—Lord Thomas Macauley, *History of England*

THE OLDEST OF SCOURGES AND THE MOST DEVASTATING

No disease has ever been so instantly recognized or so widely known and feared. Smallpox was hideous and unforgettable. For me, the memory of a ward full of smallpox victims thirty-five years ago in Dhaka, Bangladesh, is still vividly etched in my mind: anxious, pleading, pock-deformed faces. The ugly, penetrating odor of decaying flesh that hung over the ward; the hands, covered with pustules, reaching out, as people begged for help. Neither water nor food offered comfort; pus-filled lesions covered the insides of their mouths, making it painful for them to even chew or swallow. Flies were everywhere, thickly clustered over eyes half-closed by the pustules. More than half the patients were dying, and there was no drug, no treatment that we could give to help them.

Figure 1. Smallpox Deities. Sopona (*left*) was the smallpox god among the Yorubas of western Africa. **Sitala Mata** (*right*), the Hindu goddess of smallpox, shown astride a donkey, was widely worshipped in temples throughout the Indian countryside.

Dr. Nick Ward, one of my senior staff, accompanied me on the ward rounds. A veteran of medical service in Africa, he had cared for patients with the worst of tropical diseases. As we left the hospital, he placed his hands on the railing of a balcony, leaned over as he looked at the ground and said, "I don't think I can ever again walk through a ward like that. It is unimaginable." Little wonder that groups across Asia and Africa created 7 special deities such as Sopona and Sitala Mata specifically devoted to smallpox (see figure 1). No other disease warranted its own icons and in so many cultures.

Thirty years have passed since the last case of smallpox occurred. Few physicians are alive today who have seen cases outside of a textbook. Today it is impossible for anyone to comprehend what it meant to eradicate smallpox on a worldwide basis—or to envision what a devastating terrorist weapon it could be without understanding something about the disease and the virus.

A CASE OF SMALLPOX

The smallpox virus is unique among viruses in that it infects only humans—no other animals. It has survived for thousands of years by infecting one person after another in an unbroken chain of disease. Usually, transmission of the virus occurred only as a result of face-to-face contact. As soon as a patient started to develop a rash, lesions in his mouth and throat began to shed millions of microscopic virus particles into his saliva. These tiny particles would be carried into the air when he spoke or coughed. Anyone close enough to inhale them became the next link in the chain of infection.

The newly infected person felt perfectly well for the first seven to ten days. Throughout that time, however, the virus would be growing and silently establishing itself. Then it struck with the sudden onset of chills and a high fever, usually with a headache and backache so severe that the patient had to go to bed. Some people became delirious. Children sometimes had convulsions. After two or three days, the fever and symptoms temporarily ebbed. Small red spots appeared on the inside of the mouth. Angry-looking red spots cropped up on the face and, soon after, on the body; these were most dense over the face and extremities. The patient felt miserable and had trouble eating or swallowing because of lesions in the mouth and throat, which grew in size as they filled with a milky fluid and gradually became pustular. Individual pocks were buried deep in the skin and caused pain, like boils, as they expanded. There could be thousands of these pocks. Sometimes they completely covered the face, leaving scarcely a patch of untouched skin. The pustules continued to grow until nearly the end of the second week, when scabs began to replace the pustules. Among those with the severe form of smallpox found in Asia, only seven of ten unvaccinated patients survived beyond the second week.

As the scabs began to separate, symptoms disappeared and the patient was no longer contagious. Eventually, the scabs on the face went away, leaving deeply pitted scars that lasted a lifetime. Some survivors were left blind. Smallpox, in fact, was a leading cause of blindness in Europe during the seventeenth and eighteenth centuries and in India as recently as 1945 (see figure 2). All who recovered were immune for life from a second attack.

Figure 2. Blind Men with Pockmarks. Most cases of *Variola major* left facial pockmarks, which were sometimes deeply pigmented. Blindness was another possible complication of the disease. *Photograph courtesy of WHO.*

Most people had a form of the disease called "ordinary" smallpox (see plates 2 and 3). However, about one in twenty had a far more severe form, called "hemorrhagic," or "flat," smallpox, which was almost always fatal within the first week. Such patients did not develop the typical pustular lesions of smallpox, and this made diagnosis difficult. Because the rash was less distinct, these patients often infected many others before being correctly diagnosed and isolated. Another group of patients who played a significant role in transmission were those whose courses of illness were milder because of partial immunity due to previous vaccination. Having fewer symptoms, such patients could carry on many of their usual activities and would thus spread the disease to many people.

Throughout Asia, smallpox (also called *Variola major*) was uniformly severe, with a death rate of about 30 percent among the unvaccinated. In most of Africa, the proportion of those who died from the disease was somewhat smaller. In Ethiopia, South Africa, and Brazil, a mild form of smallpox (called *Variola minor* or *alastrim*) prevailed. Only 1 or 2 percent of those who developed this mild form died. In some endemic countries during the early twentieth century, both *Variola major* and *Variola minor* were concurrently present. However, during the eradication program no country had the two different forms occurring simultaneously.

THE VIRUS

The culprit of smallpox is called variola, a member of the orthopoxvirus family and one of the largest of all viruses. It consists of little more than a brick-shaped shell that houses a long strand of DNA, which carries the genetic instructions for making copies of itself. It has no means of loco-motion and is able to multiply only by invading a human cell and then taking over its metabolism and reproducing itself.

Where or how variola originated is unknown. We believe it may have started as a mutation of a related virus of the orthopoxvirus family. Such viruses affect many animals and are especially prevalent in rodents. Presumably, the strain that first infected a human changed over time as it spread and lost its ability to infect other animals. Today, humans are the only animals that can be infected with smallpox virus and that can transmit it to others.

Only three other viruses in the orthopoxvirus family can infect humans: monkeypox, cowpox, and *vaccinia*. (Chicken pox is a totally unre-lated virus.) Clinically, monkeypox looks much like smallpox. In Central Africa there are sporadic human cases and sometimes small outbreaks. The virus spreads so poorly from person to person that these outbreaks soon die out. The virus sustains itself in the tropical rain forest by spreading among small rodents. The name *monkeypox* is misleading because monkeys, in fact, are only occasionally infected. A continuing watch is being main-tained to ensure that if monkeypox ever changed so as to begin spreading rapidly among humans—like smallpox—this would be detected and stopped quickly by a vaccination campaign..

The second orthopoxvirus is cowpox, which produces skin lesions on the udders of cows and causes pustules on the skin of people who milk cows or work with them. It was a pustule on the hand of a dairymaid that enabled Dr. Edward Jenner (see plate 1) in 1796 to demonstrate that those who had recovered from a cowpox infection did not get smallpox. Over time, we have learned that the primary chain of infection of the virus is sustained by small rodents. Cows become infected from the rodents but only occasionally transmit the virus to other cows.

The third orthopoxvirus is *vaccinia*. This was the name Jenner gave to the cowpox virus material that he used for his vaccination experiments. The name comes from the Latin *vacca*, meaning "cow." The immunity provided by *vaccinia* virus protects against all of the orthopoxviruses, including monkeypox.

How long can the smallpox virus survive?

The smallpox virus's viability under differing conditions has long been of concern. As the eradication program got under way, I heard many legends that suggested the virus was exceedingly hardy: there were stories of people being infected after spending the night in a house occupied years before by a smallpox victim; of cases of smallpox developing after a long-buried corpse was exhumed; of the disease being transmitted to the recipient of a smallpox patient's mailed letter. During the program, we made special efforts to determine the source of infection of all cases—because if such legends had an element of truth, the prospects for eradication would be dim.

We paid special attention to the outbreaks of smallpox in countries that had been free of the disease for months or years: if no source could be found, the specter of possible long-term survival of the virus somewhere in the environment would be suggested. Fortunately, during the hundreds of thousands of field investigations, we were always able to identify the sources of infection with sufficient confidence so as to declare that there was no reservoir in nature.

SMALLPOX IN ANCIENT TIMES

For the smallpox virus to survive, it needs a population large enough to enable one susceptible person after another to be infected. This could not have happened until humans established agricultural communities about 14,000 BCE.

If the first forms of smallpox were as deadly as *Variola major*, the disease would have taken a heavy toll as it spread along trade routes from village to village. Throughout its history, the introduction of smallpox into

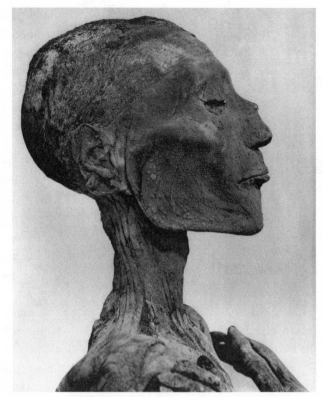

Figure 3. Ramses V of Egypt died in 1157 BCE, presumably of smallpox. His mummified head and upper torso were examined by Dr. Donald Hopkins in 1979. Characteristic lesions were present on the lower face, shoulders, and arms. *Photograph courtesy of WHO.*

new "virgin soil" populations has almost always had a catastrophic impact. However, as the disease continued to spread throughout an area, fewer and fewer susceptible people would be left to sustain the chain of transmission, and eventually the outbreak would die out until reintroduced from an infected area. When a population grew large enough so that the disease could circulate continually, the impact of an outbreak in any one year was diminished, and the virus became primarily a childhood infection.

Not surprisingly, smallpox emerged first in the early centers of urban civilization in Egypt and southern Asia. In fact, the mummified bodies of three prominent Egyptians, who died between 3,000 and 3,500 years ago, provide the earliest-known evidence of the disease: their parchment-like skin is studded with telltale pustules. The mummy of the most widely known of the three, Ramses V, is on display in the Cairo Museum in Egypt (see figure 3). Since then, the virus has claimed an unbroken chain of victims extending to the last patient, Ali Maalin, who developed smallpox on October 26, 1977, in Merca, Somalia.

In India, the first references to the disease appear in Sanskrit medical texts, written before 400 CE, where descriptions from as early as 1500 BCE are recounted. An indication that smallpox was long endemic in ancient India is the existence of the Hindu goddess of smallpox, Sitala Mata. To this day there are numerous Sitala temples throughout Hindu regions of South Asia.

The history of the spread of smallpox is fragmentary. The large populations necessary for the virus to sustain itself over long periods were present in the fertile river valleys of the Nile, the Indus, and the Yangtze. With trade, migration, and wars, smallpox emerged periodically from these endemic centers to devastate more distant populations. In the newly infected areas, many would die; the survivors would be immune, and the number of susceptible people would steadily decrease until the virus could no longer be transmitted. It would die out, sometimes for many years.

Accounts of epidemics during Greek and Roman times are sparse, but two are of special interest. The first is described as the "Plague of Athens," which began in 430 BCE and continued for two or three years, killing one-fourth of the Athenian army and significant numbers in the city. The illness was characterized by a rash of small blisters or sores, with death occurring on about the seventh to ninth day. A later series of disease epidemics, described by the Greek physician Galen, was known as the "Plague of Antonius." This devastation struck the Roman Empire in 164 CE and persisted for fifteen years—reportedly killing up to two thousand victims daily in Rome during its peak periods and causing the deaths of between three and seven million people before it ended.

The lack of adequate descriptions of the early epidemics makes it impossible to state with certainty that they were due to smallpox. Historians have speculated about other possible causes, but the events, to the extent we know them, are consistent with epidemic smallpox.

There were no other population centers in Europe comparable to those around Athens and Rome at the time of those two great plagues, and there is not much early information about European smallpox epidemics.

However, we know that in 451 CE, Hun invaders beheaded the bishop of Rheims—who was reported to have recovered from smallpox the preceding year. He was henceforth known as St. Nicaise, the patron saint of smallpox. At about this time, the Huns were compelled to retreat from Gaul and Italy because of famine and epidemics of what may have been smallpox.

SMALLPOX BECOMES ENDEMIC

As the world's population grew and spread, smallpox became endemic in many new areas. By the tenth century, the disease was an unwelcome fixture around urban areas in China, India, and Japan, in areas of southwest Asia, and along the Mediterranean coast. Traders and armies from these areas regularly carried the disease with them. This caused extensive outbreaks in more distant settlements, but then, starved of new victims, the disease would disappear from these outposts for decades. The full impact of smallpox on new, unprotected territories was particularly devastating. For example: in 1241, when smallpox first came to Iceland, 20,000 of the country's 70,000 people died.

Europe's population grew steadily from about 26 million in the eighth century to 80 million at the beginning of the fourteenth century—until 1346, when the Black Plague struck. This plague, caused by the bacterium *Yersinia pestis*, is transmitted by fleas; it wiped out between one-quarter and one-third of the population. Smallpox continued to recur. By the sixteenth century, it was well established throughout Europe, including France, Spain, Portugal, Britain, and the Netherlands—countries actively exploring the "New World" and developing colonial empires. Smallpox killed peasants and royalty alike. It supplanted plague, typhus, leprosy, and syphilis as the foremost pestilence. According to London's Bills of Mortality, which date from the mid-1600s, smallpox accounted for about 10 percent of all deaths, many of which occurred in royal families. This caused significant changes in succession to European thrones. Among those who died were Mary II of England, the last of the Tudors; Emperor Joseph I of Austria; King Luis I of Spain; Tsar Peter II of Russia; Queen Ulrika Eleonora of Sweden; and King Louis XV of France.

By the time vaccination was introduced in the early 1800s, smallpox was causing the deaths of 400,000 Europeans each year (not including those in Russia). At that time, one-third of all cases of blindness in Europe were being caused by smallpox. Separate accounts from Russia, France, and Sweden reported that at least 10 percent of all infants died each year of smallpox.

In Asia, smallpox became endemic in much of China, as well as in Burma, Siam, Japan, the Philippines, and Indonesia. Records indicate that smallpox supplanted all other pestilential diseases.

Sub-Saharan Africa was more sparsely populated than Asia or Europe, but by the twelfth century, traders from North Africa and India were regularly bringing smallpox to coastal areas. By the sixteenth century, smallpox was prevalent in tribal groups across Africa.

The only significant areas to escape smallpox were Australia, New Zealand, and the islands of the Pacific, where there were comparatively few people and contact with the endemic areas was minimal.

SMALLPOX AND THE SETTLEMENT OF THE NEW WORLD

In the sixteenth century, smallpox arrived in the Americas and proceeded to write a history of devastation unparalleled in the annals of medicine. Spanish forces first introduced smallpox in 1507. Over the next two decades, epidemics swept from Mexico to Peru, devastating the Aztecs, Mayas, and Incas—causing a higher proportion of deaths than in any other part of the world. Mortality rates of 50 to 80 percent were common. In some tribes, the survivors were so few that they could not provide for themselves. Entire tribes vanished, partly due to famine. The extraordinary mortality among the Amerindians played an important role in the expansion of European settlement throughout the Western Hemisphere even into the middle of the nineteenth century,

In Hispaniola, smallpox wiped out most of that island's population when it was first introduced. But sailors brought the disease again a decade later, and it spread quickly to Cuba and Puerto Rico, where it was said to have killed half the native population within a few months.

In November 1519, Hernando Cortes departed from Cuba for Mexico with an army of 500 men. The Aztec Empire at that time had an estimated 25 million people. Cortes's small band was not large enough to impress, let alone intimidate the emperor. However, a few months later, another small expedition headed by Panfilo de Narvaez came from Cuba. Among Narvaez's men was an infected slave who transmitted the disease to Aztec contacts. By summer, smallpox had spread to the inland plateau and from there to the capital, Tenochtitlan. The havoc is described by a Spanish friar in a history written in 1541: "When the smallpox began to attack the Indians it became so great a pestilence among them throughout the land that in most provinces more than half the population died.... They died in heaps, like bedbugs. Many others died of starvation, because, as they were all taken sick at once, they could not care for each other."

As the Aztecs prepared to drive out the Spaniards, the reigning emperor Cuitlahuac developed smallpox and died, as did many of his local chiefs. The Spaniards, survivors of childhood smallpox, were not affected. This served to reinforce the belief among the Indians that the invaders were gods. Within a year, Cortes had gathered native allies and taken possession of the capital. Two years later, one of his lieutenants invaded Mayan territory, preceded by a wave of smallpox, and then moved on through Guatemala and into the Yucatan. From Central America, smallpox spread rapidly to Peru and Ecuador and along the great Incan roads to the capital, Cuzco. The Incas were as severely afflicted as the Aztecs and Mayas. What once had been well-populated, strong, and independent societies became colonies dominated by Europeans.

Throughout the sixteenth century, the disease was repeatedly introduced into Latin America by Spanish, French, British, and Portuguese expeditions. Thus, waves of smallpox spread across the continent. No area was spared. Amerindians throughout the Caribbean appear to have been totally wiped out. In their stead were African slaves and descendants of their Spanish conquerors. On the mainland, larger groups, such as the Incas, Mayas, and Aztecs managed to survive, but it was centuries before their populations grew to pre-smallpox levels.

The effect of smallpox was no less profound in North America. The Amerindian population north of Mexico was probably not greater than 5 to 8 million in the sixteenth century. Most were hunter-gatherers or farmers. When the Pilgrims landed near Plymouth Rock in 1620, the native population was meager. It is said to have been reduced by as much as 90 percent by smallpox spreading south from a French Nova Scotian settlement two years earlier. This depopulation of the Indians was seen by many

A DOOMSDAY DISEASE

Until recently, historians have generally taken little note of the impact of disease on the Amerindian population in the settling of the Americas. Such accounts of fatal infectious disease epidemics are usually attributed to a mélange of newly introduced European diseases, such as measles, chicken pox, influenza, and smallpox. An epidemic of any of these diseases in a fully susceptible population could be serious. However, the massive number of deaths caused by smallpox was unprecedented. It came closer to final destruction of an entire civilization than any other disease, at any other time in history.

But why? A likely cause is genetic susceptibility. It was not a more virulent virus than was prevalent in Europe because, among settlers, smallpox behaved as it always had, killing perhaps 30 percent of the unvaccinated. Why were the Amerindians so seriously affected? The smallpox virus is known to have been circulating among humans for at least 3,500 years and probably was infecting humankind for thousands of years before that. With a disease so highly fatal, it would be reasonable to expect that those with a more effective immune response would be more likely to survive and reproduce. Eventually this would result in a population that was less likely to die of the disease.

Amerindian ancestors are believed to have crossed to the Americas from Asia about 25,000 years ago. Thus, it is quite possible that none had ever encountered the smallpox virus, which would not have begun spreading among humans until after establishment of the first agricultural settlements, 10,000 years later.

Europeans as a sign of God's will that they should occupy the lands of the New World. John Winthrop, the first governor of Massachusetts, wrote, "The natives, they are neere all dead of the small Poxe, so as the Lord hathe cleared out our title to what we possess." With a similar sense of divine right, the Puritan preacher Increase Mather later wrote, "The Indians began to be quarrelsome touching the Bounds of the Land which they had sold to the English; but God ended the controversy by sending the smallpox amongst the Indians."

The historian A. W. Crosby summarized the catastrophe succinctly: "During the 1630s and into the next decade, smallpox... whipsawed back and forth through the St. Lawrence-Great Lakes region, eliminating half the people of the Huron and Iroquois confederations and in 1738 smallpox destroyed half the Cherokees and in 1759 nearly half the Catawbas.... It ravaged the plains tribes shortly before... the Louisiana Purchase, killing two-thirds of the Omahas and perhaps half the population between the Missouri River and New Mexico."

This devastation continued well into the nineteenth century. In one particularly harsh seven-year period, from 1775 to 1782, smallpox swept up from Mexico, across the central plains and on into the far north. The explorer George Vancouver, cruising off the British Columbia coast in 1791, recounted discovering village after village, each big enough to have held several hundred natives. The houses had crumbled; once-worn paths were covered with weeds and scattered with human bones.

During the seventeenth and eighteenth centuries, none of the cities in North America was large enough to sustain smallpox transmission. Instead, smallpox was periodically reintroduced, resulting in large epidemics every seven to twelve years. Boston, for example, suffered five major epidemics between 1636 and 1698. But in remote and rural areas, decades might pass without a trace of smallpox. It became apparent that the longer the period of freedom from smallpox, the larger the number of vulnerable people and the more disastrous the epidemic.

EARLY PROTECTION AGAINST SMALLPOX

Until vaccination was discovered, there was no effective way to prevent smallpox other than by isolating patients and quarantining any one who might have been exposed. Traditional treatments were useless. These included herbs, noxious chemicals, bleeding, and putting patients in heated rooms so they could sweat out toxins. One such method that was unique to smallpox was erythrotherapy—the use of red color or red lights. Its traditions date back at least a thousand years and were still being employed in the early twentieth century. None of these treatments was of any discernible benefit and some undoubtedly increased the patient's risk of dying.

SMALLPOX AND THE COLOR RED

Donald Hopkins, in his remarkable history of smallpox, describes ery-throtherapy, an ancient belief through the ages in therapies based on the color red. In ancient China, physicians swabbed a victim's first pustules with red pigment and hung strips of red paper or cloth in windows and doorways. In Japan it was the practice to hang red cloths in the patient's room; small children with smallpox often dressed in red and wore red caps. Europeans, at least since the fifteenth century, also believed in the color red as a weapon against smallpox. Treatments included wrapping the patient in a red cloth, surrounding him with red blankets and red curtains, and prescribing a gargle of red wine. Elizabeth I of England was wrapped in a red blanket as a part of her care. In 1902, a physician in Iowa reported having had the windows and transoms of several rooms of a hospital covered with red paper. In the Boston Smallpox Hospital, there was a Red Room, used for especially ill patients, in which the window was covered with red curtain cloth.

In 1903, one year after Dr. N. R. Finsen received the Nobel Prize for his discovery of the efficacy of actinic rays for treating lupus, he wrote: "The action of light on the course of smallpox is astonishing, and the effect of the red light treatment is one of the most striking results known in medicine."

Variolation

Some time before the tenth century, variolation (or inoculation, as it was sometimes called) first came into use. It was the only procedure that offered any protection against the high smallpox death rates. It consisted of deliberately infecting an individual with the smallpox virus by inserting or rubbing pulverized smallpox scabs or pus into superficial scratches in the skin. Ideally, this resulted in a localized smallpox infection. In the mildest cases, a pustule—much like the pustule produced by vaccination—would begin to develop about the third day, accompanied by fever and malaise. By the twelfth day, a scab would start to form and the patient would be fully recovered and permanently protected from smallpox. That was the best-case scenario. Often, however, other pustules would appear on the arm (see plate 4)

and sometimes on other parts of the body, accompanied by more severe symptoms. In some patients, the symptoms and the rash were as serious as a typical smallpox infection. As many as 2 percent of variolated people died. Still, the risk of death was lower than the 30 percent caused by the usual method of transmission—inhaling droplets containing smallpox virus. However, a major problem for contacts and the community was that the person who had been inoculated was contagious and could spread smallpox just as easily as someone who had acquired it naturally.

The practice of variolating by the scratch method is believed to have arisen in India and later spread to China. From Asia variolation moved west along trade routes, reaching Constantinople late in the seventeenth century. Soon afterward reports of the practice reached Europe.

Variolation was not readily accepted in Europe until it was popularized by Lady Mary Wortley Montague, wife of the British ambassador to the Ottoman Empire, who learned of it in Constantinople in 1717. She herself had been seriously scarred by smallpox only a few years before. She wrote a letter to a friend in London to tell her about the process, "*Apropos* of distempers, I am going to tell you a thing that will make you wish yourself here. The small-pox, so fatal, and so general amongst us, is here entirely harmless, by the invention of ingrafting."

Lady Mary had her five-year-old son inoculated by the embassy doctor. On her return to England, she had her four-year-old daughter inoculated in the presence of physicians of the royal court in 1721. The royal family took a particular interest in the procedure and helped to promote its use in England.

The practice of variolation gradually took root, first in Britain and later on the Continent. Wider acceptance was deterred by occasional deaths from the practice and the recognition that people in contact could be infected by the inoculated subject. Objectors argued on religious grounds that only God determined who lived or died and that efforts to prevent smallpox represented a presumptuous attempt to usurp God's will.

Variolation was introduced into America at the beginning of the eighteenth century by Boston minister Cotton Mather, who is said to have learned of the practice from an African slave. Early variolation efforts in New England met with strong religious objections. To overcome this resistance, Bostonian doctor Zabdiel Boylston organized an experiment to demonstrate its effectiveness. He inoculated 300 people; only 6 of them

died. By contrast, 1,000 of the 6,000 people who acquired smallpox naturally died during the same period.

From Boston, the practice spread through the colonies and became increasingly popular. In 1775 George Washington ordered that the Continental Army be variolated. By the end of the Revolutionary War, variolation had gained general acceptance in the larger cities and towns of the United States.

JENNER'S VACCINE

Variolation was the natural precursor to the discovery of vaccination. The major difference was that in vaccination, it was material from a *cowpox* pustule—instead of from a smallpox pustule—that was scratched into the skin. The discovery, however, has been acclaimed as one of the most momentous in medical history. It was the world's first vaccine.

In 1796 Dr. Edward Jenner, an English country physician and naturalist (see plate 1), was aware of local lore claiming that milkmaids who had been infected with cowpox did not acquire smallpox. He undertook a series of practical experiments to determine whether this was true. As a first step, he took some pus from the hand of a dairymaid, Sarah Nelmes, and inoculated it into two short scratches on the arm of eight-year-old James Phipps. Three days later, the inoculation site developed redness and then a pustule. Six weeks later, Jenner variolated Phipps—and learned that he was protected against smallpox. (This was many years before the development of a medical ethics code for experimentation.)

Likewise, this was before there was knowledge that microorganisms cause disease. Today we know that Jenner's experiments were successful because the viruses that cause cowpox and smallpox are cousins. As it grows in the skin, the cowpox virus stimulates the body to produce antibodies that are similar to those made during a smallpox infection and that serve to protect against a second attack.

Jenner prepared a brief paper describing his success and submitted it to the Royal Society for publication. The president of the Royal Society, in 1797, declined to publish it on the grounds that the ideas were too revolutionary and that too little experimental work had been done. Jenner, undaunted, extended his observations by performing the same inoculation

on a number of others. A year later he published a pamphlet himself, titled (in part): *An Inquiry into the Causes and Effects of Variolae Vaccinae... known by the Name of Cow Pox. Variola vaccinae* means, literally, "smallpox of the cow," and so the procedure became known as vaccination.

Within three years, more than 100,000 people had been vaccinated in England, Jenner's pamphlet had been translated into six languages, and in 1803 a special expedition transported the vaccine to the Americas and Asia.

Vaccination was a major turning point in medical history. For the first time, it was possible to use a harmless measure to prevent a deadly disease. In 1881, in Jenner's honor, the great French microbiologist Louis Pasteur broadened the use of the term *vaccine* to refer to *any* inoculated material that produces disease immunity. Confusion has prevailed ever since. Some still use the term *vaccination* to refer only to administration of *vaccinia* virus, but most speak of vaccination as Pasteur recommended.

Vaccination was generally well received throughout Europe and was actively promoted by royalty and most clergy. In Russia, the dowager empress arranged for the vaccine to be imported from Prussia. She ordered that the first child vaccinated be named "Vaccinoff," that he be educated at

---

GETTING THE VACCINE ACROSS THE OCEAN

Cowpox was native only to Europe. Transporting Jenner's vaccine strains to Asia, Africa, and North America in the days of lengthy, precarious ocean voyages was a daunting proposition. An early and dramatic effort was made by King Carlos IV of Spain who commissioned the Balmis-Salvany Expedition (1803–1806) to transport *vaccinia* to the Americas and Asia. Twenty orphans, one of whom was vaccinated, were put on board a ship. On the eighth day, when a pustule had developed, a second orphan was vaccinated, and so on. When the ship reached its destination, the transfer of *vaccinia* was made to local residents and the vaccination chain continued.

Easier but less certain methods were also tried. One common practice was to impregnate threads with vaccine, as had been done years earlier with pustular material for variolation. Another approach was to dry the vaccine material on silver or ivory lancets. Because the poxviruses are so resistant to inactivation, enough virus often remained for a successful vaccination even after many weeks had passed.

public expense, and that he be given a pension for life. The kings of Denmark, Spain, and Prussia personally promoted the use of the smallpox vaccine. Religious authorities endorsed vaccination in Italy, Bohemia, Germany, Great Britain, and Switzerland. Some clergy not only advocated in favor of vaccination but performed the procedures themselves. The pope called it "a precious discovery which ought to be a new motive for human gratitude to Omnipotence."

Still, there were objections on theological and philosophical grounds (similar to those that had been directed at variolation). Some people argued that introducing a foreign substance into one's body was unholy and that vaccination interfered with the will of God. Rumors spread that cowpox was a venereal disease of cattle. Cartoonists drew people growing horns and tails after vaccination. One woman complained that after her daughter was vaccinated she coughed like a cow and grew hair over her body.

However, the dramatic decrease in deaths over the early decades of the nineteenth century swept away most objections (see figure 4). With vaccination, the intervals between the epidemics lengthened and the severity of the outbreaks diminished. However, major epidemics continued to sweep Europe throughout much of the nineteenth century. In part, this was because the vaccine could not be mass-produced and vaccination coverage was seldom high enough to forestall epidemics. For most of the 1800s, the source of the *vaccinia* virus was pustular or scab material from a newly vaccinated person and it was passed along by "arm-to-arm" vaccination. A week or more after a successful vaccination, pustular material or the scab was taken from the recently vaccinated person and administered to others.

The logistics of smallpox vaccination had an impact even on the American Civil War. In the North, private physicians gathered scabs from vaccinated children and sent them in individual vials to the army. Each scab was pulverized and used to vaccinate several recruits. The South, with fewer physicians to provide material, was more dependent on arm-to-arm vaccination and therefore more vulnerable to the inevitable problems of contamination. In one disastrous episode, 5,000 soldiers vaccinated with pustular material taken from a misdiagnosed syphilis patient developed primary syphilis and were too ill to take part in the Battle of Chancellorsville. Smallpox itself might have had an even greater impact had Abraham Lincoln himself not survived a bout in 1863.

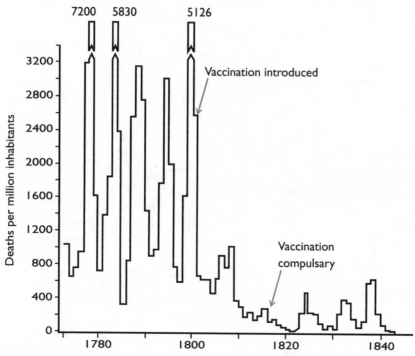

7200   5830   5126

Vaccination introduced

Vaccination
compulsary

Figure 4. Smallpox deaths in Sweden (1722–1843) before and after vaccination was introduced.

Vaccination in the United States had begun in 1800, soon after its introduction in Europe. The first vaccinations were performed by Benjamin Waterhouse, Harvard University's first professor of medicine. He vaccinated his five-year-old son and five other members of his household with material sent from England. Like Jenner, Waterhouse confirmed that the vaccine was protective by having his son variolated—the boy did not contract smallpox. As Waterhouse wrote: "One fact in such cases is worth a thousand arguments." An entrepreneur, he offered to supply the vaccine to other physicians for one-fourth of their profits or a $150 flat fee. However, others began receiving the vaccine directly from Europe, making Waterhouse's potential franchises worthless.

Waterhouse, however, played an important role as an advocate for vaccination. He engaged in active correspondence with vice president Thomas Jefferson, who himself became an outspoken supporter. Jefferson

ABRAHAM LINCOLN AND THE GETTYSBURG ADDRESS

On November 19, 1863, shortly before giving the address at Gettysburg, President Abraham Lincoln experienced a severe headache and chills. He had written the last half of his speech the night before. Lincoln returned to the White House, and two days later he began to develop the rash of smallpox. He had never been vaccinated, so far as we know. A cabinet meeting was canceled; the president remained sick in bed and received no visitors until ten days later. On December 8 he issued the Proclamation of Amnesty and Reconstruction. His personal valet, William Johnson, subsequently developed smallpox and died.

The timing in this case was crucial. If Lincoln had developed smallpox even one or two days earlier, it is possible that he might never have given the Gettysburg Address—one of the most important orations in American history. If he had not survived the smallpox, American history most certainly would have been markedly altered.

arranged for the vaccine to be made available to Indian tribes, and later sent the vaccine along with Lewis and Clark as they began their explorations. (The vaccine proved to be nonviable, unfortunately.)

The pattern of nineteenth-century smallpox in the United States paralleled that in Europe. An initial period of fairly extensive vaccination would be followed by complacency. Epidemic smallpox would recur, followed by an epidemic of vaccination, followed again by complacency, followed by yet another epidemic of smallpox. Major epidemics swept the country in 1865 to 1866, from 1871 to 1875, and from 1881 to 1883.

Establishing a program of continuing vaccination was difficult for several reasons. The logistics of arm-to-arm vaccination proved a continuing obstacle. Moreover, efforts to make vaccination compulsory inevitably generated a wave of antivaccinationist sentiment by those who in principle were opposed to mandatory regulation, and by religious groups, who argued against efforts that they believed would thwart the will of God. Furthermore, medical authorities did not yet recognize the need for periodic revaccination, and failures of vaccination undermined confidence in the vaccine's effectiveness.

NEEDED—A BETTER VACCINE

A successful vaccination program requires an adequate supply of an affordable vaccine whose potency and purity are ensured. It must also be sufficiently stable to be transported from the place where it was made to the site of vaccination activities, whether in an office or a clinic. Until the late 1800s, the *vaccinia* vaccine lacked these attributes. For nearly a century, the practice of smallpox vaccination changed little from Jenner's first experiments in 1796.

Calves become *vaccinia* factories

Arm-to-arm vaccination was cumbersome, but there was little interest in developing an alternative approach until a more serious issue arose. Vaccinators became aware that, in addition to transferring *vaccinia* virus, the organisms responsible for syphilis and serious skin infections could be transferred as well. One of the more dramatic incidents occurred in Rivalta, Italy. Use of material from a child with unrecognized syphilis spread the disease to forty-four out of sixty-three vaccinees at a children's center and resulted in the secondary infection of several mothers and nurses.

Production of vaccine by growth of *vaccinia* on the skin of a calf would seem to have been an obvious approach. However, this method was largely untried outside of Naples, Italy, where physicians began using calves as early as 1805. Because of concerns in Europe about the infections following arm-to-arm vaccination, the Italian doctors were invited to demonstrate their methodology at an 1864 Medical Congress in Lyons. They made multiple inoculations by making scratches on the calf's belly and applying material from a vaccination pustule. Seven to ten days later, they were able to take the pustular material from these lesions and use it for vaccination of humans or to inoculate other calves to produce yet more material. This method for producing *vaccinia* spread rapidly and widely. Soon, primitive "factories" were being established in a number of the larger cities around the world.

Better distribution—door-to-door cows

Transporting the *vaccinia* virus from the calf to the recipient presented an additional problem. Often, the calf was taken to a hall or other central location where people came to be vaccinated (see figure 5). In some towns, the calf was led from house to house and material was scraped off for each vaccination.

The process, although improved, clearly needed a better delivery system to permit the wider dispensing of the vaccine. Various methods evolved. The most common was to scrape the pustular material from the calf's skin on about the eighth day, add glycerol to it, strain this through a filter, and put it into very thin capillary tubes. The glycerol prevented bacteria from growing and made the vaccine somewhat more stable.

This important development permitted the vaccine to be shipped to many different sites distant from the inoculated calf. Unfortunately, the vaccine in this form could not withstand more than two to three days of

GENERAL VACCINATION-DAY AT THE PARIS ACADEMY OF MEDICINE.

Figure 5. Vaccination Day at the Paris Academy of Medicine, 1870. Material was taken directly from the cow and inoculated into the arm of the vaccinee. The use of cows for propagating vaccine was a development first made widely known in 1864. *Illustration courtesy of* Harper's.

even moderate heat. Thus, many vaccinations—especially in developing countries where refrigeration was minimal—were unsuccessful, and large numbers of people simply could not be reached for vaccination.

A heat-stable vaccine

Part of the world's failure to make better progress in controlling smallpox can be attributed to inadequate public health services and deficient infra-structures of health care, roads, and communication. However, the lack of refrigeration and absence of a dependable, heat-stable vaccine were as much to blame.

This point was vividly illustrated by the fact that Indonesia became smallpox-free for a decade in the 1930s, and several French colonies—otherwise lacking in development—also became smallpox-free in the early 1960s. A significant factor had been the use of a heat-stable, air-dried vaccine. The French and Indonesian products were not ideal; the manu-facturing methods were cumbersome and not suited for large-scale pro-duction, and the vaccine was heavily contaminated with bacteria. But both vaccines proved to be effective in the field.

The stage was set for eradication in the 1950s. Dr. Leslie Collier, working at the Lister Institute of Preventive Medicine in Great Britain (see figure 6), developed a method for producing a freeze-dried vaccine that was heat-stable for a month even when exposed to a temperature of 37°C (98.6°F). In fact, experimental batches of his vaccine retained their efficacy after being stored for more than a year at 45°C (113°F). By the late 1950s, a freeze-dryer had been perfected that could handle hundreds of vials at a time, and the United Nations Children's Fund (UNICEF) began providing them to national laboratories. When we began the global eradi-cation program, one of our first and most important tasks was to obtain enough heat-stable, freeze-dried vaccine to permit its use in all countries, even under tropical conditions.

A SECOND FORM OF SMALLPOX

A surprising development in 1897 was the appearance in Pensacola, Florida, of a very mild form of smallpox, ultimately given the name *Var-*

Figure 6. Dr. Leslie H. Collier developed a practical method for large-scale freeze-drying of *vaccinia* virus. The heat-stable vaccine proved critical for use, especially in tropical areas.

iola minor. The responsible virus is closely related to the virus that causes *Variola major*; recovery from one form protects against infection from the other. However, *Variola minor* causes many fewer deaths. Fifty-four cases occurred in 1897, but they resulted in no deaths. Within three years, this form of the disease had spread across the country and into Canada, the Caribbean Islands, and to Brazil. In 1918 it appeared in Britain and, as in the Americas, gradually displaced *Variola major*. The origin of *Variola minor* is uncertain; it was most likely from South Africa, where it had been described as early as 1895, and locally was called *amass*.

The yearly reported cases of smallpox in the United States jumped in 1900, from about 2,000 to 20,000; of these, more than 80 percent were said to be *Variola minor*. By 1920 more than 100,000 cases of smallpox were being reported annually, but the death rate was only 1 percent. In 1925 the United States reported more smallpox cases than any other country except India.

Variola minor was not considered a serious disease. As one physician reported: "[P]ersons betraying all the external evidence of the disease attended churches, schools and theaters;... officiated in public stations; and even slept in beds occupied by other non-infected members of the same family." An effort was made to have *Variola major* and *Variola minor* reported as separate diseases, but this was difficult. Many cases of *Variola minor* went unreported, and some of the milder cases of *Variola major* were indistinguishable from severe cases of *Variola minor*.

In the United States and Canada, *Variola major* continued to circulate until 1926 when, after major epidemics with high mortality in the Cleveland, Detroit, and Windsor areas, it vanished as an endemic disease. *Variola minor* cases dropped from a high of more than 100,000 cases in 1920, to fewer than 10,000 in 1933, and to fewer than 1,000 in 1942. The last cases occurred in 1949 in Texas. Success for its elimination has been attributed primarily to the increasingly widespread availability of iceboxes to preserve the vaccine, and to compulsory vaccination of school-age children.

SMALLPOX BEGINS TO LOSE GROUND

Until the mid-1920s, smallpox continued to be a serious problem worldwide, usually following a familiar pattern—epidemics of smallpox followed by epidemics of vaccination, where the vaccine was available. A few countries in Europe managed to maintain routine vaccination programs, but public health services throughout most of the world were still in their infancy, so smallpox continued to be a major problem almost everywhere. Vaccine quality control remained a stumbling block, as did shipping and preserving the vaccine—especially during hot weather. The chaos of the First World War resulted in massive epidemics in Russia, Germany, and Austria, killing an estimated 250,000 each year in Europe.

During the 1940s and 1950s, transportation and communication slowly improved and health services began to mature. Vaccine production centers developed in many countries, but most of the vaccine was not heat stable and many vaccinations were unsuccessful. Smallpox steadily decreased in the industrialized countries where the health services were the best and more potent vaccines could be ensured.

By 1959, the year the proposal for global eradication was agreed upon

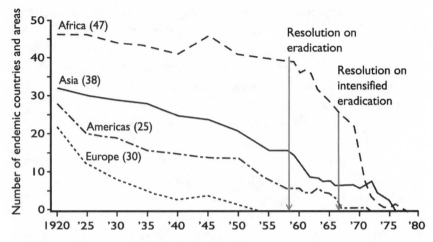

Figure 7. Countries with Endemic Smallpox (1920–1980). The number of endemic countries fell steadily from 1920 onward as vaccine became more widely available and refrigeration in some countries became possible for its storage.

by the World Health Assembly, smallpox in the Americas was endemic throughout South America, although largely confined to Brazil, Ecuador, and Colombia. Smallpox had been eliminated in Europe and most of the North African countries as well as Japan, the Philippines, Malaysia, and French Indo-China (now Cambodia, Laos, and Vietnam, respectively) (see figure 7). However, major epidemics continued to occur throughout densely populated India, Pakistan, Bangladesh, and Indonesia. Most countries in sub-Saharan Africa were effectively without control programs. In all, at least fifty-nine countries with 1.7 billion people—60 percent of the world's population—still lived in the endemic countries of the world.

Chapter 2

THE WORLD DECIDES TO ERADICATE SMALLPOX

THE BEGINNING OF THE ERADICATION SAGA

The increasingly wide availability of smallpox vaccine opened the doors to possibilities for more effective programs to cope with smallpox. By the 1950s a number of manufacturers were producing the vaccine in substantial quantities. It continued to be grown on the *vaccinia*-infected flanks of calves. However, unless refrigerated, it was usable for only a few days before it deteriorated. Thus, it was employed mainly in areas near production sites and in areas to which the vaccine could be shipped quickly. Stopping smallpox transmission throughout an entire country—much less a continental region—was a goal that few could even contemplate.

Expectations of broader goals began to change in 1950 through the efforts of a legendary public health figure, Dr. Fred Soper (see figure 8). He was director of what is now called the Pan American Health Organization (PAHO). He urged ministers of health of the countries in the Americas to join in a cooperative effort to eliminate smallpox from the entire Western Hemisphere. At the same time, he proposed regional programs to eradicate two other diseases: yaws (a syphilis-like tropical disease, primarily occurring in children and readily cured with penicillin) and

Figure 8. Drs. Fred L. Soper and Marcelino Candau. Soper, an American, was director of the Pan American Sanitary Bureau (forerunner of the Pan American Health Organization) from 1947 to 1959 and an ardent proponent of disease eradication. In 1950, he proposed programs for the eradication of smallpox, yaws, and malaria in the Americas. **Candau,** a Brazilian, was the second director-general of WHO, 1953–1973. Throughout his tenure, he was preoccupied with the global malaria eradication program. He opposed a global smallpox eradication campaign that he believed would fail and damage WHO's credibility. *Photographs courtesy of WHO.*

malaria. These were highly ambitious plans—particularly considering PAHO's minuscule budget and staff—but they were unanimously approved by the PAHO Directing Council, a group composed of health ministers of the PAHO countries.

Soper's primary expertise was in mosquito-vector control methods for yellow fever and malaria. He had not previously shown interest in controlling smallpox, much less eradicating it. But the timing for smallpox eradication was right, and as he acknowledged, his advocacy stemmed primarily from well-timed opportunism. In 1947 New York City had experienced an outbreak of twelve cases of smallpox resulting from an importation from Mexico. Fear had reigned, and 6 million people had been vaccinated. Soper learned of work in progress at the Lister Institute in England to produce a far more effective, heat-stable smallpox vaccine. He encouraged the

FRED SOPER

Fred Soper was a tall, gruff, determined man who exuded self-confidence. He believed in action. He was convinced that the technology to deal with smallpox, yaws, and yellow fever was in hand. He believed that no research was necessary and that each of these diseases could be eradicated under tight authoritarian management. Soper had already led an impressive eradication effort: during the 1920s and 1930s, working for the Rockefeller Foundation, he had directed a yellow fever eradication program throughout the Americas. This relied on destroying the breeding sites of the domestic mosquito (*Aedes aegypti*) that spread the virus. The initiative was remarkably successful, but it was thwarted in 1932 by an unexpected discovery: the virus had a reservoir in monkeys. The only possible solution was to go after the mosquito species itself.

Soper has been rightly characterized as an autocrat and a legend in his time. His yellow fever programs were quasi-militaristic with exquisitely detailed planning, tight supervision, and a veritable army of workers. However, their activities were entirely separate from public health programs. Not surprisingly, they were openly resented by many senior health staff whose best workers were regularly recruited for better-paying jobs in the yellow fever program.

There was no more ardent advocate for eradication than Soper. He argued that although eradication was initially more demanding and more costly than disease control, it would pay for itself later, because all control measures could be stopped once the disease was eliminated. Soper believed that the key to undertaking an eradication program was to obtain a national commitment to the goal and then to get on with it as quickly as possible, and to worry about strategic planning and financial resources later. He scorned research as being a waste of effort. As presumptuous as this approach may seem, his view was mirrored in 1955 as the global malaria eradication program was launched and as recently as 1988 with the decision to undertake global poliomyelitis eradication.

Michigan State Health Department Laboratories to develop such a vaccine for use in the United States and throughout the Americas. Promoting this vaccine, he believed, would demonstrate the value of PAHO as an international health agency that could offer something of value even to its biggest contributor, the United States.

For smallpox eradication, PAHO's meager resources didn't allow it to do much more than offer general encouragement to its member countries and technical support to a few national laboratories for vaccine production. Nevertheless, most of the countries did take action, using their own resources and some US bilateral assistance. By 1966, all countries in the Americas except Brazil had become smallpox-free.

1953: A GLOBAL ERADICATION PROGRAM IS PROPOSED—AND REJECTED

Meanwhile in 1953, Dr. Brock Chisholm, a Canadian—first director-general of the World Health Organization (WHO) then serving his fifth and last year—made a major proposal to the World Health Assembly. He advocated that WHO undertake a definitive program to end smallpox around the world. This program, Chisholm argued, would demonstrate "the importance WHO has for every Member State." He said that so far WHO had dealt mainly with quarantine agreements, statistical services, standardization of drugs, and the provision of advisory services to local authorities. But the organization was meant to do much more, he continued, noting that the authors of WHO's charter envisioned that the World Health Assembly would define true global programs into which country plans and requests for assistance would be incorporated. He believed that smallpox eradication was an ideal pursuit and proposed a five-year program with an annual budget of $131,000 per year.

Despite the modest amount requested, the proposal was rejected. Many delegates viewed smallpox as being a regional or even local issue. The United Kingdom delegate pointed out that the problem was vast and complicated and that "such a campaign might prove uneconomical and would not add to the prestige of the organization."

Later, committees in each of the six WHO regions discussed the proposal. But none, except for Soper's PAHO, expressed interest. The director-general sent a follow-up letter to all member states offering advice and assistance in undertaking smallpox control programs. This, too, generated little response—only two requests for consultant services—although several countries indicated they would welcome assistance with vaccine production.

Interestingly, in 1955, only two years later, the delegates to the World Health Assembly would approve a vastly more complex and costly program to eradicate malaria.

1958: THE SOVIET UNION MAKES A NEW PROPOSAL TO ERADICATE SMALLPOX

No country expressed further interest in eradicating smallpox worldwide until the 1958 World Health Assembly, held in Minneapolis (the last meeting of the assembly to be held outside of Geneva). This time, it was a Russian who brought up the subject. Dr. Viktor Zhdanov (see figure 9), deputy minister of health of the Soviet Union, presented a lengthy report about smallpox and the benefits of eradicating it globally—which he proposed be done with a four- to five-year vaccination campaign. He made no mention of the then eight-year-old smallpox eradication program in the Americas, but he took special pride in quoting a Thomas Jefferson statement of the early 1800s that supported the use of Jenner's new vaccine.

Why was the USSR concerned with the global status of smallpox? As Zhdanov later explained to me, this initiative was prompted by frequent cases coming from neighboring Asian countries and causing outbreaks in Russia's smallpox-free Central Asian Republics. He said that in the 1930s the USSR had stopped smallpox transmission nationally, even in large areas with limited health services, transportation, and communication. Conditions in some of those areas had been comparable at that time to those in most developing countries in the 1950s. The vaccine available twenty years previously had been of lower quality and was less stable when exposed to high temperatures than contemporary vaccines. Thus, he saw no reason why global eradication could not be achieved. To facilitate the effort, the USSR pledged to provide large quantities of heat-stable, freeze-dried vaccine.

It was a serious proposal made to a receptive audience. It was the Soviet Union's first appearance in the assembly after a nine-year absence from the United Nations. In welcoming the return of Soviet participation, the delegates wanted to be responsive to its proposals. Expressing agreement in principle to such a program, they asked Director-General Marcelino Candau, a Brazilian (see figure 8), to undertake a study of the

Figure 9. Dr. Viktor Zhdanov, academician and deputy minister of
health of the USSR, 1955–1960. He proposed that WHO undertake
the global eradication of smallpox. *Photograph courtesy of WHO.*

financial, administrative, and technical implications of a global smallpox
eradication effort and to report back to the 1959 assembly.

The director-general's subsequent report called for national campaigns in each country to vaccinate at least 80 percent of the population.
WHO's role would be to provide technical advice when asked and to help
in the development of vaccine production. The assembly unanimously
agreed with the proposal. But through 1966, WHO provided only modest
funding support, around $100,000 per year.

MISSION IMPOSSIBLE?

Underlying the tepid response was Candau's personal belief—shared by
many—that smallpox eradication was not possible. He believed that eradication would require vaccination of *everyone* throughout the world. And

he knew well from his own experiences in the remote areas of Brazil that it was impossible to vaccinate every single person; for example, some tribes in the Amazon seldom emerged from the jungle and simply were not accessible. A second problem was that WHO did not have the participation of all political entities. Some major countries, including China and Vietnam, were not yet members of WHO; others, such as South Africa, no longer participated or, like Angola, Mozambique, and Southern Rhodesia, were represented by colonial powers. WHO was in no position to know what each of these countries was doing in smallpox control—let alone offer technical assistance to any of them.

In 1959, the fifty-nine known endemic countries (see figure 10) reported a total of 96,571 cases of smallpox. But this was nowhere near the total of cases worldwide. Later surveys indicated that this probably represented less than 1 percent of the cases that had occurred that year. As many as ten other countries may have been experiencing endemic transmission, but the reporting was so fragmentary and incomplete that no one could be certain. Beginning in 1959 the director-general wrote annually to each country, calling attention to the new eradication program, and asking each to take action. One medical officer was recruited for WHO headquarters to oversee and coordinate the effort; however, he stayed only a few years. WHO staff members were eventually recruited to provide technical assistance as requested by four of the least-populated endemic countries: Nepal, Liberia, Afghanistan, and Mali. None of the programs made significant progress.

The Soviets were not happy. Each year at the assembly, Soviet delegates complained bitterly that WHO was ignoring the program, that except for vaccines being donated by their own country, little help was being offered. In fact, WHO was less interested in eradicating smallpox than malaria. Global malaria eradication, begun in 1955, was not going well, proving to be far more costly and less effective than had been hoped. The United States was contributing heavily to the malaria program, which also consumed at least a third of WHO's regular budget and included a staff of more than 500 people. In contrast, just a handful of paid WHO employees were working on smallpox eradication.

One other factor thought to have activated the Soviet interest was rooted in cold war politics. The United States was supporting malaria eradication, and that program was being widely publicized and discussed.

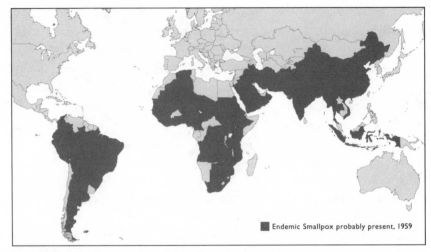

Figure 10. Endemic Countries, 1959. Fifty-nine countries
were believed to be endemic in 1959, although the
number is speculative because reporting was so poor.

Smallpox eradication, primarily supported by the Soviet Union, represented a potential counterbalance.

A number of countries joined the Soviet Union in asking the director-
general to budget more resources toward eradicating smallpox. However,
most industrialized countries—which provided two-thirds of WHO's
budget—routinely objected to any increases in the organization's budget,
even the small amounts needed to offset inflation.

Caught between the two major powers and with few discretionary
resources, the director-general was in a difficult position. He arranged for
an expert committee to meet in 1964 and advise him as to directions that
might be taken. This committee offered little advice beyond what was
already obvious. The committee noted that the reporting of cases was so
unreliable that it was difficult to assess the global situation. It recommended that countries be asked to do a better job of reporting, to use
potent vaccines, and to cooperate regionally with each other. The committee's report pointedly commented that progress was linked to available
technical and material support but offered no definitive solutions. It recommended that the director-general take all necessary steps to ensure the
success of the program. It was a useless document.

The committee did review one curious report that was provided by Dr.

K. M. Lal, an Indian member of the committee. He described an outbreak of smallpox near New Delhi that occurred despite 120 percent of the population having been vaccinated (this estimate was obtained by dividing the number of vaccinations in an area by the population). The cause for this apparent anomaly, we learned later, was that vaccinators—who had to meet quotas—repeatedly vaccinated in schools, where large numbers of children could be vaccinated quickly. Whatever the circumstance, the committee thoughtfully concluded that vaccination of as many as 80 percent of a population might not be enough to stop transmission and that the goal must be to vaccinate 100 percent. This served only to reinforce Candau's view that eradication was impossible.

The major contributors to the WHO budget, including the United States, argued that WHO funds should be used *solely* to provide technical assistance and advice, and that money for vaccines and equipment should come from voluntary donors and other United Nations agencies. Despite many requests by Candau, few donors came forward. The notable exception was the USSR, which donated large quantities of vaccine to several countries, including 450 million doses to India. Of the few donations of vaccine from other countries, half had to be rejected because they failed to meet international standards. One potential contributor of supplies and equipment was UNICEF, but it was heavily committed to supply the considerable needs of the malaria program. The director of the United Nations Children's Fund bluntly told the assembly that it "would be unable to participate in a worldwide mass eradication campaign against smallpox as it had against malaria." Meanwhile, the foundering malaria eradication program was requiring all the discretionary funds that could be mustered.

For the 1965 World Health Assembly, the director-general drew up a comprehensive report. It might be possible to eradicate smallpox within a decade, he stated, but this would take an investment in international assistance in the range of $28 to $31 million. In an extended discussion at the assembly, delegates basically repeated what had been said during previous meetings. However, this time they requested that a more detailed plan and cost estimate be drawn up by the director-general for presentation at the next assembly. Thus, the stage was set for a definitive decision on the future of smallpox eradication at the 1966 World Health Assembly. It was to mark the final act in a seven-year exercise of agonizing and hand-wringing.

THE US COMMUNICABLE DISEASE CENTER
BECOMES ENGAGED WITH SMALLPOX

While WHO was struggling with smallpox versus malaria program priorities, I was working at the Communicable Disease Center (CDC, now the Centers for Disease Control and Prevention). One of my responsibilities as chief of the surveillance section (1961–1965) was to deal with national concerns about possible smallpox importations and vaccination policies. I decided to establish a special smallpox unit to be headed by Dr. Don Millar. The unit eventually included Drs. John Neff, Tom Mack, Michael Lane, and Ron Roberto—all of whom, except for Neff, were later involved in the global program.

Why were we so concerned about smallpox? It had been more than a decade, since 1949, that the last case of smallpox had occurred in the United States. But the United States, as all other countries, continued with routine vaccination because of the fear of smallpox being imported from endemic areas and spreading widely. In the late 1950s the number of cases imported into Europe rose sharply; most of the cases originated in heavily endemic Asian countries. With air travel increasing, I assumed that it was only a matter of time before we would have smallpox importations as well. At that time, all travelers had to carry a certificate showing that they had been successfully vaccinated within the preceding three years, but this was a weak barrier; it was well known that many of these certificates were fraudulent and could be purchased easily in many countries.

In 1963 and 1964 two events reinforced our concerns. The first involved a fourteen-year-old boy who arrived in New York from Brazil and then traveled by train to his home in Toronto. He developed fever while en route, followed by a smallpox rash soon after reaching home. I flew to Toronto and quickly ascertained that the Canadian authorities had done everything necessary. They had isolated the boy, had vaccinated his contacts, and were actively seeking to find anyone else who might have had contact with him. I telephoned this reassuring news to the US surgeon general, only to learn that he was about to meet with departmental officials. He said they were on the verge of a decision to close the borders with Canada and to recommend vaccination for everyone who had been in Grand Central Station or on the train to Toronto on the same day as the boy. Some alarmists even argued that the residents of all cities on the train route should be vaccinated as well.

I pointed out that the rash had not developed until after the patient had reached Canada. Thus he would not have infected anyone en route because smallpox is not contagious until the rash develops. Did we really need to close the border, I asked—after all, if the case had been in Pennsylvania, we would not consider closing its border with New York. Why close the Canadian border? I suggested that, for public health purposes, we consider Canada to be another state. This idea was happily accepted. Later, I informed Canadian authorities that Canada was now being considered just another state. National sensitivities being what they are, this was not a good way to explain the decision. They were not amused.

The second experience, one year later, was less salutary but important in providing a reason for testing our capabilities. A woman from Ghana was hospitalized in Washington, DC, with what was diagnosed as smallpox by two Indian physicians who claimed to have seen smallpox cases and by a senior American physician who had extensive clinical experience. A specimen processed at the CDC using a newly developed fluorescence test confirmed the diagnosis. Several hundred people were identified as possible contacts of this woman; they were vaccinated and placed under a daily temperature watch so that they could be isolated immediately if they developed fever. Five days later, further laboratory studies showed that the case was not smallpox after all—merely a false alarm. Regardless, it was a useful exercise for us and resulted in the CDC's laboratory procedures being radically revamped.

Concerns about vaccine complications

Meanwhile, a growing number of physicians expressed increasingly strong objections to continuing the smallpox vaccination programs that mandated that all children be vaccinated before beginning school. The most vocal critic was Dr. Henry Kempe, a prominent pediatrician from Denver, whose hospital served as a referral center for patients with serious complications from smallpox vaccination. His frequent professional and public presentations on the subject were liberally illustrated with frightening pictures of patients with the most serious complications. Kempe insisted that such outcomes were far more frequent than anyone knew. We questioned whether they were as frequent as he argued, but we had no data. To get a better idea of the magnitude of the problem, we launched the first of two large national studies (in 1963 and 1968) under the direction of Neff and Lane. The studies showed a much lower incidence of complications than Kempe had speculated although more complications than following use of any other vaccines. However, the risk of smallpox being imported was still high enough to warrant continued routine vaccination. As the smallpox-infected countries diminished in number, it was clear that vaccination policies would have to be periodically reconsidered. Finally, in 1972, the CDC recommended that routine smallpox vaccination in the United States should be stopped.

We were especially concerned about our ability to vaccinate large numbers of people quickly should an importation occur. The vaccination technique we used in the early 1960s was the somewhat cumbersome and uncertain "multiple pressure" method. A drop of vaccine was placed on the skin, a needle lancet was held parallel to the skin, and the tip was used to press the vaccine through the skin with fifteen separate pressures. The medical teaching at the time was that if bleeding occurred, the virus would be washed out and the vaccination would be unsuccessful. It proved difficult to train physicians and nurses to press hard enough to implant the virus but not so hard as to draw blood. There were many vaccination failures, and the procedure itself was time-consuming. We needed a technological innovation.

We found it in the military, which had begun to use a jet-injector gun (see figure 11) for administering vaccines that had to be inserted subcutaneously. The nozzle was pressed against the skin and when the trigger was pulled, vaccine under high pressure penetrated the skin. It was painless and readily accepted by vaccinees. We asked the inventor, Aaron Ismach, to adapt the gun to permit the smallpox vaccine to be deposited intradermally, resulting in a more superficial inoculation. He invented a new nozzle that worked well. The vaccine was fed from a five-hundred-dose vial attached to the gun. Further refinements made it possible to operate the gun with a hydraulic foot pedal instead of electricity. In studies directed by Millar's smallpox group at the CDC and carried out in Jamaica, Brazil, Peru, and Tonga during 1964–1965, it proved possible to vaccinate as many as one thousand people per hour. Even better, nearly 100 percent of the vaccinations were successful. Although the guns cost $600 each, it

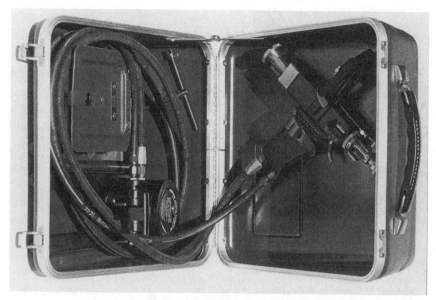

Figure 11. Ped-o-Jet Injector in its carrying case. Power was provided by a hydraulic foot pedal. The nozzle was placed against the skin; vaccine was fed from a reservoir affixed to the machine and was delivered under high pressure. *Photograph courtesy of WHO.*

seemed to be a worthwhile investment. Ultimately the jet-injector guns were widely used in Latin America, West Africa, and Zaire, but in time, they were replaced by the simpler and less costly bifurcated needle.

THE UNITED STATES OFFERS TO SUPPORT
A WEST AFRICA PROGRAM—A STARTLING DEVELOPMENT

In November 1965 President Lyndon Johnson unexpectedly announced that the United States would support a five-year program to eradicate smallpox and control measles over a contiguous bloc of eighteen (later twenty) countries in West Africa. Funding would be provided by USAID (Agency for International Development), but planning and technical assistance would come from the CDC. The objective was for equipment and teams to be in place in most countries by January 1967. This decision came as a startling surprise. Neither the CDC nor I had any experience in managing a program in a foreign country. Moreover, the CDC was then sub-stantially smaller than it is now and clearly understaffed to handle a major new activity. The US commitment, however, proved to be the final precip-itating factor that resulted in the World Health Assembly's decision to undertake a meaningful global eradication effort.

How did the West Africa program come about? It had its origins in a 1961 study of an important new measles vaccine for use in Africa. Measles at that time was a principal cause of death in African children. The study, conducted in Upper Volta (now Burkina Faso), was jointly supported by the vaccine producer Merck and USAID; it was under the direction of the Voltan health authorities and National Institutes of Health (NIH) scien-tist Dr. Harry Meyer. The measles vaccine proved to be safe and highly effective. In 1963, at the request of the national health authorities, vacci-nation was extended throughout the country to all children between six months and six years of age. Subsequently, ministers of health from five other French-speaking West African countries requested USAID support for such programs, so that their children might be vaccinated. Meyer trained a team from each of these countries. These teams, in turn, were expected to train others after the equipment had arrived in 1964. After that, further participation by NIH was considered to be inappropriate, since its mission was to conduct research and not to implement programs.

The six-country program did not go well. Personnel from some of the teams that had been instructed took other positions and were not available to train additional teams. Moreover, the program used specially designed vehicles equipped to carry six-man teams, supplies of water, a refrigerator to keep the measles vaccine cold, and other supplies. The vehicles were so heavy that they had trouble navigating in the desert. The trucks were prominently identified as Department of Public Health vehicles, but some were labeled in Spanish rather than French. Disastrously, some of these vehicles exploded and burned. It was eventually discovered that the pilot light for the kerosene refrigerator was located near the gas tank.

At the request of USAID, I agreed to send one of my staff to do a field evaluation of the status and progress of the program. I chose Dr. Lawrence Altman (who later became a respected medical writer for the *New York Times*). A six-week evaluation turned into nearly six months of work in trying to resuscitate an ill-starred array of country programs. Altman's lit-erary efforts were not always appreciated by the federal agencies concerned. In one note he pointed out that the refrigerators did not maintain a sufficiently cool temperature for the measles vaccine. USAID technical staff advised him to park the vehicles under trees. He promptly requested that they send a hundred shade trees, as he was working in the treeless Sahel Desert.

Despite such problems, there was growing enthusiasm in Washington for extending the measles vaccination program to even more French-speaking countries. For reasons I never understood, USAID did not intend to support programs in the English-speaking former colonies. It decided to continue activities in the six countries and to extend the program to three additional countries. USAID requested that I send nine staff members, each for a period of six months, to help make certain the programs went smoothly.

The basic USAID plan called for a four-year measles vaccination program in each country: 25 percent of children under age six would be vaccinated each year as teams moved quadrant by quadrant until the entire country had been covered. After the fourth year, USAID would bow out, and each country would be expected to assume responsibility for continuing its program. The extension of the program over four years was based on the time required for existing French mobile teams to reach all areas of the countries, given the limited road systems and few trained personnel. But there were two problems: first was the fact that this would not result in 100 percent vaccination coverage at the end of four years because new children were continually being added to the population pool. By the time the teams completed the fourth quadrant, all children under age three in the first quadrant would be susceptible. This concept seemed to perplex the USAID planners. The second unresolved issue was what would happen after four years. At that time, the measles vaccine cost $1.75 per dose—and these countries could not afford even ten cents per dose for the much-needed yellow fever vaccine.

I wanted to cooperate with USAID. Its resources were considerable and it was engaged in a number of disease-control programs of interest and concern to the CDC. Also, the experience acquired by our staff in participating in USAID-supported projects was invaluable. However, nine people were more than we could spare and a six-month assignment would be difficult for staff with young families. I was at a loss to come up with a reasonable compromise. Finally, partly out of desperation, I proposed an alternative plan centered around smallpox rather than measles. It called for developing a public health program that the countries themselves could support after USAID assistance ended. The plan was based on the premise that smallpox transmission throughout these countries could be stopped within four to five years—and with smallpox vaccine costing just one or two cents per dose, a continuing vaccination program was affordable. Smallpox was a serious problem in the region. If the program covered a contiguous bloc of countries, it was foreseeable that we could stop transmission throughout a large area and that this could be sustained indefinitely with country resources alone. However, there was no way for these countries alone to support the costs of a measles vaccination control program, and measles eradication was clearly out of the question. Measles is far more contagious than smallpox and, even in the United States at that time, measles continued to circulate freely.

Accordingly, the proposal to USAID was for a five-year smallpox eradication and measles control program in eighteen contiguous West African countries at a cost of $36.5 million. Included were the nine countries that USAID had already decided to support, plus all others in a contiguous bloc south of the Sahara Desert, west of Sudan, and north of Zaire. The geographic area was somewhat larger than that of the forty-eight contiguous US states. The total population was well over 100 million people. Planning called for more than 400 vehicles and 1,200 jet injectors, plus measles and smallpox vaccine, and a CDC technical advisory staff of forty-six professionals. Sixty percent of the population was in the English-speaking former colonies of Nigeria, Ghana, Sierra Leone, and the Gambia. The estimated cost was greater by a factor of five than USAID's original estimate for the nine-country program. I was convinced that the proposal would be rejected. But I thought the proposal would constitute a useful point of departure for further negotiations as to how we might address USAID's needs, given our own limitations of staff and USAID's budgetary constraints. Within a few weeks, the proposal was indeed rejected, but further discussions were planned.

At that time the White House was searching for a program that could be announced as a contribution to the United Nations observance of International Cooperation Year. At a government meeting in Washington, the Public Health Service's Office of International Health offered our proposed plan for the West Africa program. This resonated with the State Department representatives who had convened the meeting. In turn, they took the idea to the White House.

President Lyndon Johnson liked it. In September 1965, to our total surprise, we were notified of the pending approval of the complete program we had proposed. Dr. Alex Langmuir, my chief, was furious (see figure 12). He pointed out that our epidemiology branch staff had no experience in running an international program of any type—much less one of this magnitude. I pointed out that I had not expected that any more than a semblance of the proposal could conceivably be approved. Langmuir refused to have the epidemiology branch play a role in the program. Instead, he requested that all of the proposed activities be assigned to a special branch reporting to the director of the CDC. The confrontation was brutal and remains a vivid memory. As he said: "Get out. Take whatever staff you believe you must have. Move off this floor. I want nothing to do with this operation." We had little further personal contact for more than a year. Gradually, our relationship warmed, and Langmuir became a strong supporter of the program. He served on several national certification commissions.

It was a number of years before I came to appreciate more fully Langmuir's response. I had been working closely with him for the best part of nine years—formative years for myself—for the two-year training program called the Epidemic Intelligence Service (EIS) and for the CDC epidemiology program. Alex and I shared many confidences, and responsibility for the branch often fell to me when he was away. He valued greatly his personal contact with each of the EIS officers. He was an outstanding teacher and sought to have each officer write at least one paper for publication. For each of these papers, he labored long hours in critiquing and mentoring. As the EIS grew, he became perceptibly more concerned about having less time and opportunity to spend with the individual recruits. The expansion of staff anticipated for West Africa and the new administrative load would have completely swamped the epidemiology branch staff and its chief.

It was the end of an era for me, but the West Africa program gave an unexpected impetus to the faltering global commitment to eradicating smallpox. The US delegates to the World Health Assembly were instructed to pledge support for an international program "to eradicate smallpox completely from the earth within the next decade."

Figure 12. Dr. Alexander Langmuir, chief of the epidemiology branch of the CDC, 1949–1970. He introduced me to public health and epidemiology during my service with him from 1955 to 1966. Langmuir developed the concept of disease surveillance, which was the foundation for the critical surveillance-containment strategy of smallpox eradication. *Photograph courtesy of the CDC.*

THE DIRECTOR-GENERAL CHALLENGES THE 1966 ASSEMBLY

Director-General Candau forced the 1966 World Health Assembly to decide whether or not it was serious about eradicating smallpox. The acid test was whether the delegates were prepared to allocate enough money from the regular budget. For seven years delegates had complained that the director-general was not devoting sufficient resources to the program, but he could respond only by diverting sparse budget funds from other programs. The director-general had solicited additional support from other United Nations agencies and possible donors but had received few contributions. Understandably, Candau was distressed by the growing criticism of himself and the organization for failure to make better progress. Moreover, it was apparent that there were a number of senior scientists who were skeptical about the very concept of eradication—of any disease.

The 1965 assembly had requested that Candau draw up a plan with detailed costs for eradicating smallpox over a ten-year period. In the autumn of that year, I was asked to work with WHO staff in Geneva in developing these estimates. This had been difficult. We had no cost data about current vaccination programs and little information as to the extent and severity of smallpox throughout the world. Nevertheless, we managed to draw up a thirty-seven-page document, setting forth a proposed strategy, time lines, and expenses. This was distributed to all countries in

THE FUTILITY OF ERADICATION

The widely read and respected medical scientist, Dr. René Dubos, took a dim view of the concept of eradication and wrote about it in 1965 in his book *Man Adapting*. It was a view that was widely shared at the time. However, by the time I got around to reading the book, my future had already been decided.

At first sight, the decision to eradicate certain microbial diseases appears to constitute but one more step forward in the development of the control policies initiated by the great sanitarians of the nineteenth century. . . . In reality, however, eradication involves a new biological philosophy. . . . Social considerations make it probably useless to discuss the theoretical flaws and technical difficulties of eradication programs, because more earthly factors will certainly bring them soon to a gentle and silent death. Eradication programs will eventually become a curiosity item on library shelves, just as have all social utopias.

April 1966 and discussed at the World Health Assembly. We estimated that there were about 1.1 billion people in the endemic areas and that we would need to vaccinate between 200 and 350 million people each year over the next ten years. Based on a crude approximation of ten cents per vaccination, the overall cost would be at least $180 million. For the first year, we foresaw a requirement of $22 million; 70 percent of this was expected to be borne by the countries themselves. We forecast that the balance of $6.6 million would have to come from international agencies, voluntary contributions, and bilateral assistance. Of this amount, Candau decided that WHO could provide no more than $2.4 million—and even this he could not do within his current budget without crippling other vital activities.

Candau was almost certain that a request for a budget large enough to accommodate the additional $2.4 million for smallpox eradication would be rejected. Thus, he decided to present two budgets to the assembly—one that included special funds for smallpox eradication and one that did not. It was implied in a statement to the delegates: "If you are as serious about smallpox eradication as you claim, then provide the organization with the special funds I have asked for and let us get on with the task. Otherwise, accept the fact that your expectations of progress will not be met and the responsibility for not being able to do so will rest with the assembly." It was not an unreasonable charge.

A new WHO budget was considered annually and any increase was a point of contention. The regular budget consisted primarily of annual assessments of each of the member countries. Each contributed an amount roughly proportionate to its gross national product. About two-thirds of the total was provided by the five largest economies—and these were usually the ones who objected most strenuously to the increases. Not surprisingly, the debate as to whether or not to pursue smallpox eradication came down primarily to a question of money.

Candau's budget without provision for smallpox eradication was $49.2 million, which was 11.5 percent higher than the year before—an increase that was almost entirely the effect of inflation. If the delegates desired to provide support for smallpox eradication, then a budget of $51.6 million was required—an overall increase of nearly 16 percent. If the larger proposal was rejected, smallpox eradication could continue but with little expectation of success for the foreseeable future. To Candau this was an acceptable outcome. He believed that smallpox eradication was not achievable in any case and that the failure of such a program would discredit WHO—which, with a failing malaria eradication effort, was already being heavily criticized.

The debate was lengthy. As expected, many countries objected in principle to the proposed increases in budget. Others questioned the feasibility of smallpox eradication itself and some expressed doubts about launching a new eradication program at a time when the flagging malaria eradication campaign needed all the help it could get. However, the fact that the United States had recently committed to providing all necessary support for smallpox eradication in eighteen West African countries was strong, tangible affirmation of support for the active pursuit of global smallpox eradication.

The debate lasted for three days, after which both budgets were put to a vote. A total of fifty-eight votes was required for approval of the special budget; sixty votes were cast in favor, twenty countries voted against, and twelve abstained. This approval by just two votes was the narrowest margin for any action in the history of the assembly. Although no records are kept as to how individual countries voted, I was told by a US delegate that its vote reflected its own ambivalence. Most of those on the US delegation were enthusiastically supportive of eradication and lobbied for the special budget. (To put this in perspective, the US share of the proposed increase would be $800,000.) However, the delegation had been instructed by the State Department to vote against the budget if it provided for the additional funds for smallpox eradication. And so they voted against the budget they had lobbied to support.

Candau was displeased that the special budget had been approved and mainly blamed the United States, whose advocacy, he believed, had been a critical factor in the vote. He insisted to the US surgeon general William Stewart that an American had to direct the program so that, when it failed, the United States would be seen to bear major responsibility for its approval. Specifically, Candau requested that I be given this job.

Soon afterward, Assistant Surgeon General James Watt called me to Washington to inform me that I was being assigned to WHO in Geneva. I declined—citing, in particular, my considerable new responsibilities as director for the recently approved West Africa program. That program, I pointed out, was crucial to determining the feasibility of global eradication. If smallpox transmission could be stopped in these African countries —the world's most heavily infected and with the least resources—it would provide strong encouragement to programs throughout the world. I agreed with others that the prospects for global eradication were not bright, given that only $2.4 million in WHO budget funds would be available annually. This would provide an average of less than $50,000 a year for each of the fifty endemic and neighboring countries where special programs were needed. It was not even sufficient to purchase the vaccine needed for the program. Voluntary contributions amounting to several times the budgeted amount were projected, but there had been little response to solicitations in the past. Finally, WHO had little experience, except with the malaria program, in developing and operating collaborative international programs. Moreover, I was well aware that the six very independent WHO regional offices were unaccustomed to working closely with headquarters staff or with each other.

My arguments were in vain. Watt told me that he had no choice but to order me to assume the directorship of the program for at least eighteen months in order to get it launched. He said that if I found the position to be untenable, to simply send a telex stating "Now," and he would bring me back.

The Henderson family—myself, my wife, and three children ranging in age from six to eleven years—put half of its household goods in storage and moved to Geneva. We were not to see the furniture again for eleven years.

Chapter 3

CREATING A GLOBAL PROGRAM

A PROGRAM IN ITS INFANCY

On October 26, 1966, I arrived in Geneva to face the stark realities inherent in assuming the position of chief of the Smallpox Eradication Unit—to direct a global program intended to reach more than 1 billion people in fifty countries. I was thirty-eight years old and had a mere ten years of public health experience. Many thought I looked considerably younger than thirty-eight; certainly I lacked the maturity and gravitas expected of a WHO unit chief, few of whom were then under fifty.

I was struck by the disparity between the extravagant expectations and the smallpox unit's modest, three-room headquarters; its very small staff; its shoestring budget; and few other resources. My new office, a room of about 200 square feet, was equipped with one bare desk, a chair, a telephone, a side table, and a metal book shelf. But the view was spectacular. We were on the sixth floor of a modern building overlooking the city of Geneva, the Alps, and Lac Leman (see figure 13). Unfortunately, we didn't have much time to enjoy the scenery. Problem after problem arose—most of them demanding precedent-setting solutions. We were beginning from scratch.

The unit had four other staff: Dr. Stephen Falkland—on temporary loan from another program—a debonair, grizzled veteran of international

Figure 13. WHO Headquarters, Geneva. *Photograph courtesy of WHO/J. Germain.*

health programs; two secretaries; and Dr. Isao Arita, a dedicated, capable Japanese physician. Arita was about my own age. He had recently been transferred to Geneva after a year as the WHO smallpox adviser in Liberia. We were to work closely together for the next eleven years, at which time he assumed direction of the program and carried it through to certification. Eventually, the unit's space would be doubled, and we would add a medical officer, two administrative officers, and another secretary. Even then, considering the scope of the activities and the desired results, nine people was a skeleton crew. My pleas for additional space and staff, however, were summarily rejected.

For our mission to succeed, every country would have to participate: special programs of some sort would be needed for all countries where smallpox was endemic as well as those adjacent to them. We would need to persuade governments to undertake programs in which the governments themselves were expected to provide at least 70 percent of the national costs. Global, regional, and country-specific planning meetings would have to be convened. We would need training programs, operations manuals, a network of vaccine production laboratories and testing facilities, and new national and international surveillance programs. Because two-thirds of the projected budget for international support was expected to come from

WHY WE BELIEVED SMALLPOX COULD BE ERADICATED

Smallpox and the freeze-dried smallpox vaccine possessed unusual characteristics. Taken together, they were unique and made smallpox, by far, the best available candidate disease for eradication. Most important was that humans were the only victims of the smallpox virus; there was no reservoir in nature. No rodents, monkeys, or other animals could be infected. Each person who was infected exhibited a rash that could be identified even by illiterate villagers. No laboratory tests were required. If patients were promptly isolated, they could be prevented from spreading infection. By contrast, most infectious agents (for example, polio, tuberculosis, hepatitis) cause many subclinical infections as the disease spreads silently in a population. Moreover, a patient could infect others only during the two to three weeks of severe illness; on recovery, the person was immune for life.

The development of freeze-dried vaccine was a critical advance. It could withstand storage at 98°F (37°C) for a month, making it ideal for tropical areas. The vaccine was inexpensive, vaccination was easily performed, and a single vaccination provided immunity for at least ten years. Every successful vaccination resulted in a pustule and then a distinctive scar, which remained for decades. In areas where *Variola major* had been the prevalent form of smallpox, 80 percent of those who recovered had permanent scars. Thus, teams visiting an area could readily determine whether smallpox was present in the community, when it had occurred in the past, and who had been successfully vaccinated. No other disease came close to being such an ideal target.

voluntary contributions, we would have to spend at least some time in the delicate art of fund-raising. Balancing the timing for all of this would be crucial: if one country's program lagged or failed, other countries would be at risk of importing new cases.

In many ways we constituted a unique type of program at WHO. Although the malaria eradication division was coping with similar challenges, it had been operating for more than a decade, its budget was many times larger, and it functioned as an all-but-independent entity within WHO. In the countries in which the program was in place, there was little contact between its personnel and those in other health programs. Most other WHO units served primarily to advise governments when requested, to compile occasional reports for the annual assembly, and to deal with such issues as quarantine and standardization of drugs. They had no direct responsibility for the operation or performance of national programs.

The structure of WHO's headquarters organization was a disjointed mélange of programs and activities spread across more than eighty different units and offices, many with no more than two or three people. They had sprung up like weeds during WHO's first twenty years of growth. About half of these had administrative and liaison functions; about forty dealt with issues of public health. The smallpox unit was one of the forty. The public health units were overseen by six division directors and two assistant directors-general (ADGs). I was told not to expect much contact with either our division director or our ADG, as they were largely engaged in representational and liaison functions. In a report ten years later, I estimated that I had spent no more than ten to twelve hours per year with my division director and perhaps one to two hours per year with my ADG. For the first three years, however, we were fortunate to have a division director, Dr. Karel Raska from Czechoslovakia, who was a strong advocate for smallpox eradication. He was a dour man whom we saw infrequently but who was one of the few senior officials who believed that smallpox eradication was possible. He provided valuable help in recruiting young Czech epidemiologists for the program and sometimes intervened with our ADG to speed up the often indecisive decision-making process.

I had been told that most questions of policy and program would be taken up directly by Director-General Candau or his deputy with individual unit chiefs. For some units, contacts of this sort were said to be fairly frequent, but for me they were rare. This was not surprising, considering Candau's antipathy toward the concept of smallpox eradication. For day-to-day policy and program decisions, and for dealings with regional directors and national leaders, we were effectively on our own.

I needed help to navigate this tangled bureaucracy. One of my earliest acts was to find a capable administrative officer. The ADG for administration, Milton Siegel, a US citizen, helped in many ways but especially in identifying an extraordinary administrative officer, John Copland (see figure 14). He was a paragon among an important group of program operations officers. These staff members made up for a lack of formal public health training with intelligence, common sense, flexibility, and imagina-

Figure 14. WHO Geneva Smallpox Eradication Staff. Long-term members:
(*A*): D. A. Henderson, chief, 1966–1977; (*B*): Isao Arita, medical
officer, 1965–1976, chief, 1977–1985; (*C*): John Copland, administrative officer, 1967–1977; (*D*): John Wickett, technical officer,
administrative officer, 1970–1987; (*E*): Susan Woolnough, secretary,
1970–1985; (*F*): Celia Sands Hatfield, secretary, 1969–1981.
Photograph courtesy of WHO.

tion. Some had unconventional backgrounds that made them unlikely candidates for the usual international health program. Copland received a Yale degree in musicology, spent three years as a US Marine Corps officer, and undertook graduate studies in Paris. He came to WHO as a malaria program administrative officer in East Pakistan and West Africa. He was brought back to Geneva because of his familiarity with the WHO bureaucracy, but he quickly went beyond the prescribed spheres of budgets and job descriptions. He crafted an office staff whose priorities were to provide immediate support to those in the field and who could assist in any area, from finding a spare part to keep a vehicle on the road to aiding in a personal crisis of one of the field staff.

Copland was joined three years later by an equally energetic and imaginative young Canadian, John Wickett, whose background in public health was even less impressive than Copland's. He had graduated from the University of British Columbia with a degree in mathematics, roamed around Europe for a year, and taught math and skiing. He joined us in 1970 to work out a computer program for recording smallpox case data, rapidly self-schooled himself in WHO bureaucracy and smallpox, and stayed to work in smallpox activities for the next eighteen years. With Arita, Copland, and Wickett, plus two British secretaries, Sue Woolnough and Celia Sands, we had a dedicated, stable core group, a critical need for an overworked, ever-traveling professional staff.

An especially commendable role was played by the WHO administrative staff. They increasingly went out of their way to support our efforts—creatively implementing means to help bridge financial crises, to streamline recruitment, to arrange for emergency shipments of vaccine, and even to annually restore our totally depleted cash reserves by feats of accounting sleight-of-hand. We regularly invited them to join us at unit social gatherings, but, as I was to learn, there was a strange cultural chasm between WHO professional staff and administrators. To illustrate, my ADG, Dr. P. M. Kaul, was angry that I had asked Siegel for help in recruiting an administrative officer. He said that he didn't want a spy from administration working in one of his divisions. By the time Kaul learned of this, however, Copland had already been posted to the unit and nothing could be done to rectify this perceived grievous error.

COUNTRIES, FIEFDOMS, AND
SHORT-CIRCUITING THE BUREAUCRACY

The regional offices of WHO were important components of the administrative structure. For smallpox eradication, they were more a hindrance than a help. All WHO member countries belonged to one of the organization's six regions. In 1967, four of the regions included countries with known endemic smallpox. One regional smallpox eradication program adviser was allotted for each of three—Southeast Asia (SEARO), Eastern Mediterranean (EMRO), and the Americas (PAHO). Two advisers were allotted for the African Region (AFRO)—one for eastern Africa, based in Kenya, and one for western Africa, based in Liberia. The advisers were selected and appointed by each regional director without reference to our unit at headquarters.

The regional directors considered themselves all but autonomous. After Dr. Halfdan Mahler became director-general in 1973, he often said to me, "You have to remember that WHO is, in fact, an Association of Regional Offices, not a *World* Health Organization." The regional directors were each elected for four-year terms by a majority vote of that region's member countries. Gaining or retaining a country's vote required skillful politics, and such factors inevitably played a role in important decisions such as the selection of qualified staff and allocating of budget funds.

The regional directors interpreted the initiatives of the World Health Assembly as being primarily advisory. Some they accepted, some they ignored, and some they modified to suit their own and the region's particular needs and agendas. Given the director-general's openly expressed skepticism about smallpox eradication, it was not surprising that there was little support for smallpox eradication in the regions. Two regional directors were passively or openly hostile to the program; one largely ignored it; and one actively opposed it. Fortunately, the one who most strongly opposed the eradication program was replaced through election by a strong supporter a year after we began.

The regional offices added a dysfunctional layer of bureaucracy for communications with country program directors and WHO advisers. Recall that at the time of the program, there were no e-mail facilities, no mobile phones, and no fax machines. Telex and ordinary telephone calls were expensive (and not always technically possible). Personal contact and correspondence by mail were our only reliable routes for communication. WHO policy required, however, that all correspondence with countries had to pass through the regional offices. But this was more than simply a routing of memos. For example, if I wanted to communicate with a smallpox program adviser in Uganda, my letter had to go first to the African Regional Office whose office was in Brazzaville, Republic of the Congo. It would be read by one of the staff, who might eventually prepare a draft forwarding memo for the regional director's signature—a deceptively simple procedure that often required several rewrites and could take weeks. The recipient, the WHO country representative, would then consult with our WHO smallpox eradication adviser. A reply followed this track in reverse. This bureaucratic tangle ensured that four to five months might elapse between the time I sent a simple query and received a reply—if one came at all. Eventually I resolved the problem by simply sending original copies of memos to the regional office, as directed, and carbon copies directly to the recipient. However, this added another wrinkle because the WHO mail pouch did not carry personal mail and the copies were so regarded. Regular postal service worked reasonably well, although it often cost me 100 to 150 francs a month to send my documents. It was worth it.

The direct system of communication sometimes created its own problems: once we received a telex from Uganda asking that 2 million doses of vaccine be sent urgently. We sent the vaccine by air the following morning. Five months later, the regional office wrote to report an urgent request from Uganda for 2 million doses of vaccine. We did not know whether this was a new request or the one we had dealt with five months previously. (It was the latter.) The policy of quietly short-circuiting the regional office, when necessary, continued for years and, surprisingly, was never questioned, if indeed the regional directors ever learned about the practice.

THE CREATION OF THE PROGRAM BUDGET

Smallpox eradication was allotted $2.4 million each year. I had assumed, wrongly, that I would have substantial input as to how this might be strategically distributed. That never happened.

Each year the director-general submitted a budget to the annual assembly for review and for routine formal approval. It was detailed down to country and individual program, showing the amount of money and the number and rank of WHO personnel to be assigned. As a first step, he decided on how much to allot to each region. Next, the distribution by country was decided by the regional directors. I never saw the allocations in our $2.4 million smallpox eradication budget until it was presented to the assembly as a printed document.

The assembly "debated" the budget, made suggestions, and offered general observations but seldom altered the allocations. After the budget was approved, the regional directors had flexibility to shift funds from one program to another and even from country to country. As director of an ostensibly integrated global program, my only hope to institute such a change in allocations was to persuade a regional director to do so. This proved all but impossible. On the one occasion I was successful, the agreements I negotiated were subsequently ignored.

In 1967, on the grounds that smallpox eradication was impossible and a waste of time, effort, and money, SEARO regional director Dr. Chandra Mani decided not to spend most of his smallpox funds. With difficulty I persuaded him to take the unorthodox step of transferring this money to PAHO, where we hoped to bring smallpox quickly to an end in Brazil— thus ending the disease in all of the Americas. Then we could concentrate our funds in Africa and Asia. I assured Mani that when smallpox had been eradicated in the Americas, compensatory allocations would be assigned to SEARO. Five years later, with the Americas free of smallpox and money desperately needed for the program in India, I asked PAHO regional director Abraham Horwitz to agree to transfer funds to SEARO. He flatly refused. When I appealed to the director-general, he shrugged his shoulders and reminded me that he was, after all, only the director-general.

HOW MANY SMALLPOX CASES?

We were well aware that smallpox cases were seriously underreported and guessed that we were being notified of about one case in twenty (5 percent) at most. For example, when 118,000 smallpox cases were reported in 1967, we estimated that the true number was probably closer to 2 million

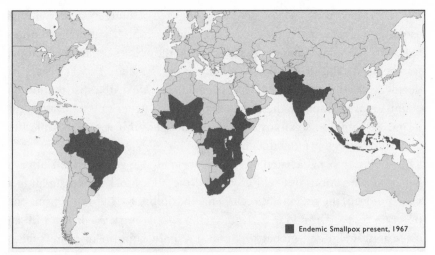

Endemic Smallpox present, 1967

Figure 15. Endemic Countries—Beginning of Eradication Program, 1967. Thirty-one countries were endemic; twelve others experienced imported cases.

cases. Later surveys revealed that probably fewer than 1 percent of cases were being reported. Thus, as the program got under way, the number of smallpox cases in forty-three countries was probably between 10 million and 15 million with 2 million deaths. Of these countries, thirty-one had endemic smallpox; in the remaining twelve countries, smallpox had been imported. The endemic countries included most countries in sub-Saharan Africa; the Asian countries of India, Pakistan, Afghanistan, Nepal, and Indonesia; and Brazil (see figure 15). More than one billion people were living in the endemic countries.

It was encouraging to discover that the Western Hemisphere, except for Brazil, was probably free of smallpox. So, too, were a number of populous Asian countries including Thailand, Burma, Korea, the Philippines, and Vietnam. China had stopped smallpox transmission in the early 1960s, although this was not known to us until the late 1970s. Two factors common to the successful countries were reasonably well managed: community-wide vaccination programs and the use of a heat-stable vaccine. None of the programs had endeavored to build an effective case-reporting system, however, and thus we could only speculate as to which countries were truly without cases. Definitive information had to await the development of effective surveillance systems.

THE REALITIES OF EXECUTING A SIMPLE, TWO-PART STRATEGY

Our original strategy was straightforward and consisted of two basic components: the first was preventive—vaccination to reduce the number of people who were susceptible to smallpox so as to diminish spread; the second was to organize a reporting-surveillance system and response teams to investigate and contain outbreaks. Implementation would have to adapt to national realities. Each country had its own health system and each was unique. Other differences had to be taken into account—the quality of personnel and the transportation system, religion and politics, climate conditions, and demographics. There was no possibility for a detailed plan that could be used everywhere. Creativity and flexibility in every program were not only welcome, they were to be encouraged and successful new concepts communicated to others.

Mass-vaccination strategy

Except for the South Asian countries of India and Pakistan, vaccination coverage was thought to be sufficiently poor as to require mass-vaccination programs. The overall plan for such programs called for campaigns to cover the entire country systematically, village by village, in three years or less and to reach at least 80 percent of the population. Why 80 percent? Some speculated that this percentage was chosen as a level of "herd immunity" at which smallpox transmission would be expected to cease. In fact, the figure was arbitrary, based on what veteran international staff deemed to be an optimal achievement. We discovered that 90 percent coverage was actually feasible where good support from community leaders had been obtained. Resistance to vaccination was not expected to be significant, and, except for a few areas in Africa, it was not.

Many WHO health policy experts were hostile to the use of mobile vaccination teams. They argued that all vaccinations should be provided through existing health centers, to build up the "basic health services." However, we found health centers in most endemic countries to be few and far between, usually understaffed, poorly supervised, and preoccupied with dispensing drugs and treatments. Where only local health centers and hospitals were used for vaccination, it was difficult to obtain a coverage of more than about 60 percent. Mobile teams proved to be highly cost-efficient and more effective. In Africa, a six-vaccinator team using bifurcated needles was expected to vaccinate an average of 3,000 people per day, and as many as 600,000 over the course of a year.

Special quality-control teams, each with two or three people, were assigned to visit a sample group of villages, ten to fourteen days after the teams had vaccinated. They checked to ensure that at least 80 percent had been vaccinated and that at least 90 percent of those vaccinations had been successful. If an area was not completely covered, the teams would return and revaccinate.

Surveillance and containment—a new component

Prior to 1967, smallpox control and elimination programs consisted solely of mass vaccination. The new component that we added called for the development of a national reporting system to obtain weekly reports of cases of smallpox from all health units as well as teams that could go to infected areas to vaccinate and so control outbreaks. The concept would seem all too simple and obvious to warrant being singled out as a vital change in the strategy of smallpox control. However, it was a new approach for public health staff in all countries and, like most new ideas, was initially resisted by many.

The 1964 WHO Expert Committee on Smallpox made no mention of surveillance. It merely pointed out that reporting was so incomplete that no conclusions could be reached about the status of smallpox. In 1965 I was asked to go to Geneva to help develop the director-general's report to the 19th World Health Assembly. I argued the case for surveillance and containment of outbreaks being an essential component. Having spent the preceding five years as chief of the surveillance section at the CDC, I had been deeply involved in the development of surveillance systems for a number of different diseases. To me, it seemed logical that surveillance should be a key part of the new smallpox initiative.

From the director-general's report to the 19th World Health Assembly, March 28, 1966:

> It is necessary for the eradication programs to develop a systematic plan
> for the detection of possible cases and the concurrent investigation

regarding the source and site of disease acquisition, their vaccination status and the prompt instigation of containment measures. Detailed epidemiological investigation of all cases as to the reasons for their occurrence and the means by which they are being spread can be one of the most effective instruments to provide continuing guidance and direction in the vaccination program.... An outbreak, however small, demands a full critical review with appropriate revisions of the program.... Even in countries with a limited local health structure, a systematic surveillance plan can and must be developed as an essential component of the eradication program.

The change in strategy proved to be a critical one for many national programs. The first to implement the strategy, in January 1967, was a CDC team in Eastern Nigeria, headed by Dr. William Foege. They had arrived in Nigeria before most of their supplies and had limited amounts of vaccine and equipment. They decided to use the vaccine in focused vaccination programs in areas near the sites of active cases. Mobilizing a missionary radio communications system for reporting, they rapidly developed a surveillance system for the region and gave priority for vaccination to areas where the most cases were occurring. At the end of six months, they had recorded 754 cases and vaccinated some 750,000 of the twelve million people in the region. But smallpox transmission had been interrupted. In September 1968 the CDC introduced this approach (called "Eradication-Escalation") into the other countries of the West Africa program. Meanwhile, a similar experience in interrupting smallpox transmission was reported from Madras State in India in June 1968 and in Parana State, Brazil, in 1969. Gradually, surveillance-containment began to be extended to other countries—but it was a slow process.

The basic inputs for surveillance were weekly reports from all health centers and hospitals, indicating the number of cases of smallpox. This information was to be compiled concurrently and analyzed, and the reports disseminated widely. Progress in finding cases and reducing their number was to be the primary measurement of progress instead of the old yardstick—the number of people vaccinated. To strengthen this component, two- or three-person containment teams were created. They would respond quickly to investigate reported outbreaks, verify cases, search for additional cases, and vaccinate contacts and others in the

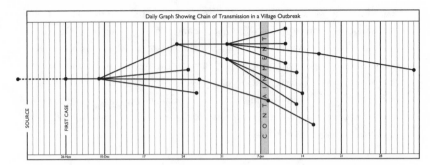

Figure 16. Smallpox Surveillance Diagram, an illustration of the standard chart used for tracing chains of infection during field investigation. Cases were entered on the grid by date of onset. A line was drawn between cases to depict how the disease spread. This type of chart served to encourage field staff to investigate more carefully each outbreak. One case seldom infected more than two to five others.

immediate area—creating a barrier that would block further transmission of the virus (see figure 16).

Once the surveillance-containment methodology was put into operation, the effect in stopping smallpox transmission was often dramatic, even in poorly vaccinated areas. For every case investigated, the teams regularly found at least twenty other cases; in Brazil, it was more often a hundred for every case, and in Ethiopia, nearly a thousand. Because of better reporting, the numbers of reported cases often rose sharply at the beginning of a campaign. Often this proved to be advantageous since the increase was frequently interpreted at high levels of government as the advent of a serious epidemic. Alarmed officials took a heightened interest in the program and provided additional resources. Sometimes, however, as in both India and Bangladesh, the effect was the reverse, as orders were given to stop the surveillance-containment activities and to assign immediately all available staff for mass vaccination.

AVAILABILITY OF VACCINE—
AN APPARENTLY SOLUBLE PROBLEM IS ANYTHING BUT

The technological breakthrough that made smallpox eradication possible was a stable, freeze-dried vaccine that would not require refrigerated storage even in tropical countries. Providing adequate supplies of freeze-dried vaccine that met international standards was a top priority.

In 1967 we knew that the vaccine was being produced in seventy-seven different laboratories. But how good was it? That we did not know. We projected that we would need at least 200 million doses of vaccine each year. The cost of the vaccine, even at one cent per dose ($.01), would have consumed our total budget. The only possible solution was to rely entirely on locally produced or donated vaccine.

In the beginning I had not anticipated that the supply of vaccine would be a major problem. The Union of Soviet Socialist Republics (USSR) had pledged to donate 25 million doses annually, and the United States was to provide the 50 million doses each year for the twenty countries of West Africa. Several other countries had pledged donations as well, although they had not specified how much. In addition, three of the largest countries—India, Indonesia, and Brazil—were reported to be producing quantities of vaccine that I assumed were sufficient for their own needs.

As a first step, we insisted that all vaccine for the program be tested by independent laboratories for potency, purity, and stability. The deputy director-general pointed out that WHO had never monitored the quality of vaccines produced by national laboratories and had no authority to do so. Authority or not, I simply advised countries that WHO could support only programs that used vaccine that met WHO international standards. Directors of laboratories in the Netherlands (Rijks Institute) and Canada (Connaught Laboratories) agreed to do the testing. We found that national authorities generally supported our insistence on vaccine testing, but their laboratory directors did not always agree. As we soon learned, few laboratories were adequately testing their vaccine, and those with the poorest vaccines protested the loudest about submitting specimens.

To our surprise and dismay, we discovered that only about 10 percent of the vaccine being used met accepted standards. Some samples had no detectable vaccine virus at all. Meanwhile, we found that few producers of satisfactory vaccine had a capacity of more than a few million doses per year. Thus, hopes of receiving substantial additional vaccine donations from the industrialized countries evaporated.

We had no choice but to launch an urgent program to improve vaccine production in the endemic countries. Arita convened an emergency consultation, bringing in experts from five major laboratories in the United

States, USSR, Canada, Great Britain, and the Netherlands. They developed a detailed, step-by-step manual of standard production and testing procedures, and later made site visits to many laboratories to help put these into operation. The United Nations Children's Fund (UNICEF) provided needed production equipment. We advised countries with small laboratories and those in need of extensive renovation to close down—with the promise that WHO would supply national needs. By 1973 more than 80 percent of the vaccine in use was being produced in the recently endemic countries, and all vaccine met international standards.

Until 1967 the handling of donated vaccine had been the responsibility of a different unit in WHO. Our unit assumed this burden—for good reason. The practice had been that when a country offered to donate vaccine, WHO thanked the donor and waited until a needy country asked for vaccine before requesting samples. The rationale was to make sure that the donated product would be satisfactory before it was shipped out. Test samples were then sent to a laboratory in Denmark that processed specimens only two or three times a year. If the vaccine proved to be satisfactory—and only then—WHO asked the producer to send the vaccine directly to the requesting country. The lengthy procedure meant that between twelve and eighteen months frequently elapsed between the time a country requested vaccine and the time it was actually received. We changed the routine. When a laboratory offered a donation, we immediately requested samples and tested them promptly. If the specimens met international standards, we had the vaccine shipped to Geneva where we stored it in a rented, refrigerated warehouse. Our goal was to be able to respond within twenty-four hours to any reasonable request for vaccine.

VACCINATION TECHNIQUES—THERE HAVE TO BE BETTER WAYS

The vaccine—its quality and quantity—was one hurdle. Administering it was another. The techniques in use were not ideal. Since Jenner's day, doctors had used a number of different techniques for implanting the *vaccinia* virus into the superficial skin layers where it could multiply and cause a protective infection. Most vaccinations were performed by placing a drop of vaccine on the skin and making multiple scratches through the drop, using a small lancet. In the United States, a needle was placed parallel to the skin and the virus was pressed lightly beneath the skin using the tip. This technique, called multiple pressure vaccination, was less traumatic than previous methods, but there were many vaccination failures. In Southeast Asia, the most common method used was a "rotary lancet" (see figure 17). Five short prongs protruded from a small metal plate. The plate was pressed firmly into the skin over the drop of vaccine and given a short twist. Four of these were performed on the forearm or upper arm. The insertions were painful and often became infected. In areas where these were in use—primarily India—resistance to vaccination was not un-common, and some officials wondered why.

In the early 1960s investigators at the US Communicable Disease Center (CDC) experimented with a jet-injector gun. As many as one thousand people could be vaccinated in one hour. Unfortunately, the guns—although high-performance—broke down frequently so that a trained technician had to travel with each team; one or two backup guns were needed for each one in operation.

In 1967, soon after the global program began, we discovered the ultimate vaccination solution: the bifurcated needle, invented by a Wyeth Laboratory scientist, Dr. Benjamin Rubin (see figure 18). Made of stainless steel, it was two inches (five cm) long, with a two-pronged fork at the end. The bifurcated needle was intended for a multiple pressure vaccination, but I suggested that we try holding it at right angles to the skin and making fifteen rapid punctures to the skin (see figure 19). This was simpler, and because of the fork, the needle could not penetrate too deeply. A trace of blood usually appeared at the site after fifteen to twenty seconds. Medical textbooks at that time warned that if bleeding occurred, the vaccine virus would wash out and the vaccination would be unsuccessful. The opposite proved true. Nearly every vaccination with the new "multiple puncture" method was successful—better still, most local vaccinators could learn the proper technique within ten to fifteen minutes. A plastic needle dispenser (see figure 17) was designed in Pakistan by Dr. Ehsan Shafa, an Iranian, who was our WHO regional smallpox adviser. With the needle, we could vaccinate a hundred people with a vial that previously (using the older techniques) had provided only twenty-five vaccinations. Suddenly, our ever-stressed vaccine supply was quadrupled. Best of all, the needles could be boiled and reused almost indefinitely—and they cost only US $5 per thousand.

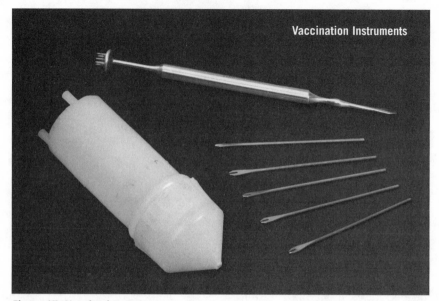

Vaccination Instruments

Figure 17. Vaccination Instruments: Rotary lancet—Used primarily in India. The end with the prongs was pressed over a drop of vaccine on the forearm and twisted. A slightly longer central prong served to hold the lancet in place while twisting. Four insertions were usually made. **Bifurcated needles**—First used in 1968. They replaced all other instruments. **Plastic holder for needles**—One needle at a time was shaken out for vaccination. Used needles were placed in a second holder. To sterilize, the holder with used needles was immersed in boiling water for twenty minutes. When finished, the water was shaken out through multiple small holes in the bottom of the holder. *Photograph by Will Kirk/Johns Hopkins University*

As a last step in simplifying vaccination, I questioned the need for using alcohol to cleanse the vaccination site. A 1960s study in the *British Medical Journal* reported that wiping the vaccination site with alcohol did little more than rearrange the bacteria. Therefore, we did away with alcohol and cotton swabs and directed that if the site was caked with dirt, it should be wiped off with a damp cloth.

Simplicity and speed were achieved. The only pieces of equipment a vaccinator required were the now hundred-dose vials of freeze-dried vaccine, vials of a saline solution to reconstitute the freeze-dried vaccine powder, a pot for boiling the needles every night, and two plastic tubes, one for clean needles and one for used needles.

Figure 18. Dr. Benjamin A. Rubin, a scientist at Wyeth Laboratories. He was the inventor of the bifurcated needle, a simple device as revolutionary as the safety pin. *Photograph courtesy of Wyeth Laboratories.*

COMMUNICATION—CONFLICT AND CONTROVERSY

Another important task was keeping governments interested, supportive, and aware of both progress and problems—and at the same time keeping in touch with smallpox program directors and our WHO staff. Until 1967 little was heard about smallpox eradication except at meetings of the World Health Assembly. My plans for raising awareness included ensuring that smallpox eradication was on the agenda at each WHO Assembly, issuing periodic up-to-date surveillance reports on numbers of smallpox cases and where they were occurring, and providing reports of progress and problems. Distribution of the surveillance reports would be to the widest possible audience—smallpox staff, health and government officials, and the public press.

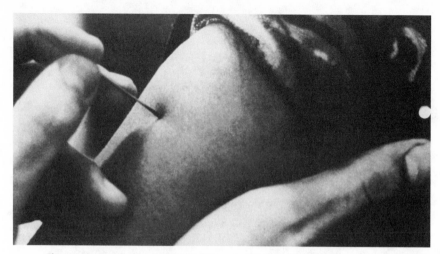

Figure 19. Vaccination with the Bifurcated Needle. The needle is dipped into the vial of vaccine and withdrawn; vaccine is held between the prongs by capillarity. With the needle at a right angle to the skin, fifteen rapid strokes are made within a small area. The procedure is less painful than an injection given with a needle and syringe. *Photograph courtesy of WHO.*

These initiatives seemed straightforward and noncontroversial—and yet problems I did not foresee made them difficult to implement. A major roadblock, as mentioned earlier, was Candau's certainty that smallpox eradication was destined to fail. He had lobbied hard against taking on smallpox eradication in the first place. The next best thing was to give the program the least possible visibility—and to hope that when it collapsed, it would attract the least possible attention.

Off and on the World Health Assembly agendas

The World Health Assembly met each May for three weeks to address a number of health programs, a variety of political issues, and budget allocations. Each country was represented by its minister of health or equivalent. The director-general drafted the agenda, but additional items could be added at the request of members of the assembly's executive board.

When smallpox eradication was on the agenda, as in 1964, 1965, and 1966, many different delegations requested time to tell what their countries were doing and what they needed, as well as to raise questions. As many as sixty or seventy interventions—questions or comments—might be made. This was called a "debate," after which resolutions were passed. But the primary importance of the discussions, as I came to realize, was that if the subject was on the agenda, ministers took an interest in the subject. They sought information from their staffs about their own programs, reviewed what was being done (sometimes worrying about facing public criticism if their country was not doing well), and took action accordingly. Thus, to have smallpox eradication on the agenda was our one opportunity each year to capture the attention of ministers.

After the first year, smallpox eradication was regularly omitted from Candau's proposed agenda. However, because board members could ask for an item to be placed on the agenda, the subject was added annually either by the USSR or the United States. Additional help was provided at each assembly by Soviet and US delegates who would visit me before the meeting and ask what questions I would like them to raise that might highlight difficulties as to why some country programs were faring poorly or what administrative barriers our program was encountering.

Three reporting systems become one

It was important to document for health officials what was being accomplished. Regular reports of numbers of cases by country provided a useful overall indicator. Until 1967, however, such data were viewed as so woefully incomplete as to be almost meaningless. Thus, from the beginning, we made a concerted effort to obtain prompt weekly reports from every country. We also introduced the phrase "Best Available Data" (unfortunately, the acronym is BAD) to characterize the tabulations.

We soon discovered that from many countries we had at WHO received two and sometimes three different sets of numbers, each tabulated by different offices. One set came weekly from quarantine officers to the WHO International Quarantine Unit. Another data set was collected by WHO statisticians for an annual publication on the occurrence of diseases in the different countries. These data were often compiled by statistical bureaus separate from health departments. Finally, Ministry of Health infectious disease offices obtained data through health officials at state and district levels.

None of the data provided a current or reliable picture of progress. The quarantine data consisted primarily of lists of districts or counties that had reported cases of smallpox in the preceding weeks. The numbers of cases were of little interest to those collecting or compiling the information. Data for the annual publication were obtained yearly and usually published two to three years later. It took nearly three years to persuade governments, and WHO, to assign the duty for smallpox data collection and verification to those directly responsible for eradicating smallpox. At WHO, our unit became the sole responsible repository. We pressed for more rapid reporting at every level and harassed authorities that were delayed in their submissions. Reporting improved steadily thereafter.

A surveillance report threatens the whole program

I needed to be able to communicate frequently and effectively with program directors and WHO smallpox staff scattered across the globe—to chart progress and to talk about challenges, setbacks, and new observations. At the CDC, for each disease with a surveillance program, we had published a regular report distributed to program staff, state epidemiologists, and others with public health responsibilities. It seemed logical to undertake a similar publication at WHO.

We published the first *Surveillance Report* in September 1967. In it we summarized our just-completed review of the sorry status of smallpox vaccine quality and urged more testing. We tabulated reported cases of smallpox by country and pointed out trends and challenges. It went out to about two hundred people. We printed our second report in December and sent it to the WHO mailing office for distribution. In late January, I received a phone call informing me that the report could not be sent. The director-general had decided that WHO was distributing too many different documents and wanted a committee to do a study before anything more was published. I was told the study would take a year. I made an impassioned plea to the chairman of the review committee, who, after several more weeks, finally gave an approval to send it out—provided that I redid the report so that it had a different appearance. I asked what it was that needed to be changed, but he did not seem to know. I did a hasty rewrite. A month later I was informed that the amended version could not be distributed.

I was at a loss as to what to do, but finally I decided on a course of action that I had never before taken—to resign my position. I met with

Director-General Candau and informed him that I was incapable of operating the program if I had no means of communicating with our many geographically scattered programs and staff. I suggested that he appoint someone who could. He asked for time and eventually made an alternate proposal—that the summaries be published periodically in WHO's *Weekly Epidemiological Record* (*WER*). This was a weekly publication that listed every county (or similar unit) of every country where designated quarantinable diseases had been reported. It made for very dull reading, and I doubted that there were many who really paid much attention to it. However, it was routinely sent to five thousand people, and Candau agreed that a special distribution could be arranged for smallpox program directors and the WHO staff. I could not have been happier. By May 1968 the first of our reports was published. These continued at three-week intervals through 1979. Eventually, the *WER* began to publish other substantive reports and eventually evolved into a well-read publication.

DIPLOMATIC CHALLENGES—
THE COLD WAR AND OTHER PROBLEMS

Even during some of the darkest days of the cold war, our eradication program continued to function without apparent restraint as an international entity with a responsibility for facilitating cooperation across political and cultural barriers. An illustration of this was my own position. I had worried as to what the attitude of the USSR might be toward a program director from the United States and, in consequence, toward the program itself. I had learned that the Soviet Union expected a Russian to be named director. After all, it was the USSR that had proposed the global eradication program, and was one of its most ardent supporters. The Soviets had committed to provide 25 million doses of freeze-dried vaccine each year—a quantity far beyond the proposed donations of all other countries combined except for the United States. Without the USSR's commitment, the program would be in deep trouble.

I introduced myself to Soviet vice minister Dimitri Venediktov (see figure 20) at the World Health Assembly meeting soon after I arrived and expressed my concerns. He walked me to a quiet corner and said, "We have checked you out, and have favorable reports of your performance at CDC.

More than that, we believe that your only goal is that of eradicating smallpox. You have our full support." As for the donation of 25 million doses of vaccine every year, he laughed and said I should not be concerned. He was certain that I didn't understand the socialist system—that once a large amount of vaccine such as this begins to be produced each year, it is impossible to turn off the tap.

The extent of cooperation was tested a year later. I received a memo from our WHO adviser in Afghanistan, who enclosed a USSR lab test that showed that a Soviet gift of vaccine they had received did not meet international potency standards. I wanted to go to Moscow to discuss the problem, but WHO's deputy director-general refused to approve the trip. He argued that WHO must not involve itself in relationships between two countries, especially when the USSR was involved. I waited a couple of months and finally received approval to visit—although my stated purpose was for other reasons. I immediately went to Venediktov's office and explained the vaccine problem. He listened carefully and apologized. He said, "I will promise you that no smallpox vaccine will ever again leave the USSR that does not conform to international standards." He kept his word.

Recruiting Russians was an especially sensitive issue at WHO. The USSR believed that there should be proportionately more Russians in the organization than there were. Each year, the Russian delegates to the assembly brought many applications for individuals who were said to be qualified, requesting their appointment. Some were qualified; some were on the list because of political influence. Positions were found for many who were ill-suited for the responsibility assigned to them and, as might be expected, widespread dissatisfaction ensued.

It was important for the smallpox program that we have competent young epidemiologists who were reasonably fluent in English and willing to work under difficult circumstances. I was confident that there were a number of such individuals in the Soviet Union but did not know how to identify them. I turned to Venediktov, who proposed that I fly to Moscow and, with him, interview candidates for possible appointment. We jointly interviewed some thirty to thirty-five young physician-epidemiologists, about half of whom looked promising to both of us. In due course, many joined the program and performed well.

Figure 20. Dr. Dimitri Venediktov, deputy minister of health for the USSR, member of the executive board of WHO, and strong supporter of the program. *Photograph courtesy of WHO.*

YET ANOTHER PROBLEM—OBTAINING A COMPETENT STAFF

Screening and appointing new staff for regional and country posts were responsibilities about which I was seldom consulted and over which I had no control. When the program began, many staff were transferred to smallpox eradication from other assignments where their performance had been less than remarkable. During my time in US government service, I had learned that a newly created program inevitably inherited a number of staff whom supervisors desired to replace. Smallpox eradication was no exception. Perhaps a third of those who had been transferred responded well. Many remained deskbound and disinterested. They were most distressed by my stipulation that all professional staff must spend at least one-third of their time working in the field. Some resigned and some moved on to other posts.

Eventually, our smallpox eradication professional staff, excluding short-term consultants, came to number about sixty-five people. At head-quarters, we never had more than nine; in the four regional offices combined, five; and at the country level, between fifty and fifty-five.

During the final four years of the program, we encountered desperately difficult problems in India, Bangladesh, Ethiopia, and Somalia that required additional effective field staff. This situation was eventually solved by recruiting three-month short-term consultants and extending them if they proved satisfactory and wanted to stay. We looked for young professionals who were intelligent, motivated, able to work well with people, and willing to put up with rugged field conditions. Whether they knew anything about smallpox was irrelevant; they could learn to diagnose smallpox and the program strategies after they joined. Dr. David Sencer, then chief of the CDC (see figure 21), provided many able people—including veterans of the West Africa program. Others were referred by senior staff of schools and institutes who knew our program well and could personally identify suitable candidates. Recruits included many from Sweden, Poland, USSR, Czechoslovakia, Netherlands, United Kingdom, France, Switzerland, and Canada. As countries such as Indonesia, Brazil, and those in Africa became smallpox-free, we recruited former national program staff. In all, some 765 individuals from seventy-three different countries served as WHO staff or consultants at some time during the program. However, the total number of international staff in the field was never large; at the peak of activity, in 1974–1975, we had a maximum of 150 international staff, including volunteer contingents of young people from the United Kingdom, Japan, Austria, and the US Peace Corps.

Given the small number of WHO staff for a program of this magnitude, it is apparent that the foundation for the program and the responsibility for its ultimate success were attributable to national personnel, from ministers of health to program directors to field supervisors, vaccinators, and laboratory and administrative staff. In the mid-1970s, as many as 150,000 workers were engaged. A number of them sacrificed other careers and opportunities, and some even gave their lives.

Figure 21. Dr. David Sencer, director of the US Communicable Disease Center, which oversaw the CDC-supported West Africa program and dispatched many CDC staff to critical areas. *Photograph courtesy of CDC.*

RESULTS OF A TRANSFORMATION

Our first few years were especially rocky. For smallpox eradication to succeed, many of WHO's traditional policies and procedures had to be interpreted more flexibly—or, in some cases, changed. An effective, cooperative international effort of a unique character had to be generated. WHO's active involvement in the execution of national public health programs had to be accepted. Quality control measures for vaccines, vaccination, and surveillance containment mechanisms had to be developed and applied.

After the first several years and some early program successes, fewer difficulties began to be encountered in charting new directions. In part, this reflected a more supportive director-general, Dr. Halfdan Mahler, who took office in July 1973. Mahler, a veteran of national tuberculosis vaccine (BCG) campaigns, understood many of the problems we were facing. He was personally supportive of smallpox eradication and made a number of critical interventions to help resolve pressing problems that we were facing. However, the ultimate success of the program remained uncertain through the autumn of 1977 and the containment of the last outbreak.

Chapter 4

WHERE TO BEGIN?

A Tale of Two Countries— Brazil and Indonesia

Our limited WHO budget would not allow us to support more than a few of the endemic countries during the first years of the program. I believed that we would need to set priorities, to focus at first on directing more resources to a few countries. With those programs satisfactorily under way, we could gradually shift the funds to other countries. The concept made sense and seemed simple. Putting it into action was another story.

At WHO, there was surprisingly little appreciation of the fact that smallpox eradication would require very different approaches from those used in other WHO-assisted programs. A collaborative global effort was needed with strategic planning and budgeting, an internationally coordinated reporting scheme, and regularly updated accounts of progress. No other program was comparable in scale or in its requirements. But then, no other program had been charged with executing a global program requiring participation of all countries.

Smallpox eradication was a square peg in a round administrative hole. Existing WHO programs made few efforts to coordinate national programs or to oversee the performance of national programs. At the time, WHO's country activities were diverse—separately initiated products of national requests and regional office decisions. The projects tended to be small, providing little more than the salary of a special WHO adviser or consultant plus travel and meeting costs. Our particular ideas of strategies and plans, of targets, priorities, and accountability were not welcomed, nor were some of our creative solutions. Some were openly resented as being impingements on the authority of the regional directors. Not infrequently, it was made apparent that we were considered an annoying, irrelevant nuisance. However, we persisted in our efforts to see a logical plan unfold.

In 1967 more than fifty countries either had endemic smallpox or were adjacent to countries where smallpox was endemic. One of the most problematic areas was the twenty-country group that constituted West Africa. Inhabited by more than 100 million people, some of this region's countries were the world's most heavily infected and seriously underdeveloped. Fortunately, in November 1965, the US government had agreed to make available $35 million in support of a five-year program to eradicate smallpox and to control measles. The Communicable Disease Center (CDC) was given responsibility for management, technical guidance, and support. I had been responsible for the development and planning of this program during the year prior to my assuming direction of the global program. By March 1967 the West Africa program was well under way. In sixteen of the countries, plans, resources, and advisory personnel from the CDC were in place. Programs would be extended to the remaining four countries in 1968.

My concern now was how to proceed in developing needed programs in the thirty or so other countries whose population numbered more than 1 billion people. More than half of that population resided in India, which was already fully engaged in executing a massive eradication campaign. Additional contributions from our modest budget would be of marginal importance.

Of the other endemic countries, two of the most populous offered strategic opportunities—Brazil and Indonesia. Each was far enough from other endemic countries so that, once they were smallpox-free, the importation of new cases was unlikely. To focus on these seemed to be an obvious policy choice. This was not to be. Decisions about priorities were firmly held by the WHO regional directors, each of whom could choose to support or even ignore special initiatives. For the most part, they supported programs set forth in resolutions of the World Health Assembly. Smallpox eradication was not one of these. The negative reactions of the regional directors reflected, in part, Director-General Candau's view that smallpox eradication was impossible. In fact, he suggested in a memo to them that, should they wish, the money allocated for smallpox eradication could be used for almost any purpose that could be deemed to strengthen the basic health services in some way—in brief, virtually any program qualified for the funds.

STONEWALLING

The problems of developing and implementing programs in Indonesia and Brazil exemplified the obstacles. Dr. Chandra Mani, WHO's regional director for Southeast Asia (SEARO), blocked an initiative in Indonesia for almost a year. Mani had decided that smallpox eradication in Asia was a lost cause, and that the funds allocated to his region for smallpox eradication would be wasted. Thus, he refused to permit his staff to visit or even communicate with Indonesian health officials.

Fortunately, I was able to persuade Mani to transfer much of the money that had been allocated to SEARO to the Pan American Health Organization (PAHO), to bolster the Brazilian program. Brazil accounted for half of the population in South America, and interrupting smallpox transmission appeared to be a fairly straightforward task. The country was reporting only 3,000 to 4,000 cases each year and it had a moderately well-developed infrastructure of transportation, communication, and health services. Once Brazil was free of smallpox, it was unlikely to be reinfected since no other Latin American country was reporting smallpox cases. However, interrupting smallpox transmission in this country of fewer than 100 million people took more than four years—substantially longer than we had anticipated. There were several reasons: regrettably ineffectual leadership in PAHO, indifferent advisory staff assigned to Brazil by PAHO, frequent changes in the national government's political and public health leadership, and the failure of Brazilian laboratories to produce a fully potent, heat-stable vaccine.

The programs in the two countries presented completely different problems, but both programs demonstrated the need for an effective working relationship between headquarters, the regions, and the countries—and accountable leadership at every level of responsibility. The situation aptly illustrated the need at that time for a truly *World* Health Organization.

Figure 22. Brazil population (1970): 95,900,000.
South America population (1970): 192,500,000.

THE BRAZILIAN PROGRAM—A REGRETTABLE SAGA

Between 1950—when the PAHO Directing Council voted to eradicate smallpox—and 1967, the Americas had made surprising progress. The United Nations Children's Fund (UNICEF) had participated in developing freeze-dried smallpox vaccine production laboratories in nine countries. By 1967 eight of them were producing vaccine that met or were close to meeting international standards. The one exception was Brazil.

During the early 1960s, most national governments in South America conducted periodic large-scale vaccination programs, often in response to epidemics. The numbers of reported smallpox cases dropped steadily until, in 1966, the last endemic cases outside of Brazil were recorded in Argentina and Peru. However encouraging this may have been, we remained uncertain as to the true situation in South America, because nothing had been done to improve a seriously deficient reporting system.

Brazil continued to report a few thousand cases each year, but it was certain that there were, in actuality, many times that number. *Variola minor* (called alastrim) had displaced the far more virulent *Variola major* and was now the only form of smallpox. It was considered to be no more serious than chickenpox. Many patients had comparatively few symptoms, and children who were ill sometimes were able to attend school. Not surprisingly, officials were less concerned about the disease than were authorities in countries where many died from smallpox.

The best strategy, or so I had advocated, was to use the bulk of WHO funds to support community-wide, systematic vaccination campaigns throughout Brazil and, at the same time, to develop the surveillance reporting systems throughout the continent. However, the idea of surveillance was dismissed out of hand. Most of those in leadership positions— in PAHO and in Brazil—had earned their status through performance in the field during the extensive malaria eradication campaigns that had begun in 1955. The primary strategy of the malaria program had been mosquito vector control. Only after this had been achieved were efforts made to find and treat people with malaria. The mindset for smallpox eradication, on the part of both national and international staff, was the same: first, national mass-vaccination programs, then, only afterward, development of a reporting system to find the last remaining outbreaks of smallpox and to stop transmission.

The regional director, Dr. Abraham Horwitz, distributed the allocated PAHO smallpox funds as well as the funds transferred from SEARO to eleven different countries—to support further mass-vaccination campaigns. Less than half of the region's funds went to Brazil. The decision had its political overtones—to provide all countries with a share of the newly appropriated smallpox funds, whether needed or not. Nothing was said or done about surveillance for smallpox cases. The regional office appointed a full-time regional adviser for smallpox, Dr. Bichat Rodrigues, a Brazilian. Three physicians were recruited to serve as smallpox advisers in Brazil, all having been transferred from other PAHO posts. None had experience with vaccination programs, only one spoke Portuguese, and none exhibited much interest in eradicating smallpox.

The single encouraging element in the picture was PAHO's agreement to recruit a statistician-epidemiologist. Candidates were scarce, so I volunteered the services of Leo Morris (see figure 23), who had been my right-hand man in the CDC Surveillance Unit. Although the most junior of the advisory group, Morris took full responsibility for restructuring Brazil's dysfunctional case reporting system, in training Brazilian counterparts, and in developing epidemiological field studies. At the time of Morris's arrival, only half of Brazil's states and territories were sending any reports at all. One of his first projects was the creation of a monthly surveillance bulletin. It had a profound impact on the program. The first issue was published just five months after the program began. It was distributed to more than one thousand senior health and program staff plus schools of public health and medicine. By documenting cases by week and by state, the program put pressure on those who did not report. The bulletin also included analyses on trends of the disease, schedules of the mass-vaccination campaign, and the results of special field investigations. It instructed, motivated, and gave a widely scattered staff a sense of common purpose.

Figure 23. Leo Morris, statistician-epidemiologist, was assigned from the CDC to the Brazil smallpox eradication program (1967–1970). *Photograph courtesy of CDC.*

Mass-vaccination programs began in 1967 with expectations of vaccinating 30 million people each year. But supervision was poor, and in the first year only 6 million were said to have been vaccinated. Even these figures had to be questioned. One of the smaller states completed its vaccination program by March of the first year, but three months later, twenty-one cases were reported from a town of 6,317. A vaccination team reported that it had vaccinated 6,558 residents of the town over a six-day period. If an outbreak of this size occurred in a thoroughly well-vaccinated town, something had to be wrong. Morris and his team undertook a special house-to-house investigation. They found thirty unvaccinated cases and discovered that only 49 percent of the village had actually been vaccinated. It was clear that some sort of quality control system was needed to check on the vaccination teams. Until this study was done, the national program director had refused to have some teams "do nothing but evaluate." The director was terminated and a new director appointed: Dr. Oswaldo da Silva, an experienced former malaria program manager. Four-person teams were established in each state to regularly undertake sample evaluations of areas that had been vaccinated seven days earlier. Where the teams functioned, the quality of coverage improved, reaching 90 percent among children under fifteen years. Overall progress, however, was disappointing. National funding support fluctuated, and the vaccination program failed to reach its targeted momentum until 1970.

Vaccine problems

A continuing, worrisome problem was the quality of the vaccine. The regional office had made arrangements for Canada's Connaught Laboratories, a government-owned production laboratory in Toronto, to work with vaccine laboratories in Central and South America, testing the quality of samples and visiting regularly for on-site consultation. The concept was excellent, and the quality and quantity of vaccine in most national laboratories rapidly improved. However, the Brazilian vaccine consistently failed to meet standards for potency and stability. The vaccine is supposed to contain at least 100 million vaccine virus particles per milliliter, a high enough concentration to ensure that more than 95 percent of vaccinees would be protected. The vaccine was also expected to maintain full potency when kept at 37°C (98°F) for one month.

Part of the problem in the Brazilian laboratory had to do with the culture medium. Most of the vaccine was grown on egg embryos rather than on calves. In principle, this was a good thing because it resulted in a more pure vaccine. But the vaccine was less potent and far more susceptible to being inactivated by heat. Because of the frequency of vaccination failures, it was certain that at least some batches of vaccine did not meet standards, but the extent of the problem was difficult to verify; comparatively few samples were sent from Brazil to Connaught Laboratories for independent tests. My recommendations to PAHO to arrange for extended, on-site consultations by Connaught scientists were rejected by the regional office. The stated reasons were concern about national pride and qualms about offending the scientists. The result was that the program was conducted with a fragile, unstable vaccine with demonstrably poorer successful vaccination rates than in other countries.

Surveillance-containment saves the day

The development of a functional surveillance-containment program throughout the country was a principal need. Mass-vaccination programs reduced the numbers of susceptible people and numbers of cases; however, early detection of outbreaks and vaccination of contacts was needed to stop smallpox spread and was essential for confirming eradication. Regional Adviser Rodrigues simply smiled and agreed whenever I talked with him about surveillance-containment, but he did nothing to implement it. Fortunately, I was asked to attend a meeting in Brazil late in 1968 and while there had the opportunity to meet with the general secretary of the Brazilian Ministry of Health, Dr. Nelson Morais. He was uninformed as to what a surveillance program was. I provided descriptions of the successes of surveillance-containment in West Africa and India and proposed that he assign one medical epidemiologist to each state's department of health. He reluctantly agreed. Morris promptly organized a course in surveillance-containment methodology. It was attended by thirteen medical officers working in state mass-vaccination campaigns plus three recently graduated epidemiologists from schools of public health. Despite what had been agreed with the general secretary, the national smallpox program director remained committed to mass vaccination. He directed that the medical officers begin surveillance programs in their states—but only after their vaccination campaigns had been completed.

The three epidemiologists from the schools of public health were assigned to work in four states where vaccination campaigns had not yet begun. Because surveillance was such a low priority, only one was given a car and driver. The others would have to use a state vehicle, if one was available, or else travel by bus. Despite less than ideal working conditions, these epidemiologists—Drs. Ciro de Quadros, Nilton Arnt, and Eduardo Costa—provided startling and unexpected data. They investigated twenty-seven reported cases—and discovered thirty-three separate outbreaks with an additional 1,465 cases. This information was duly reported in the national bulletin and widely publicized throughout the country. The number of reported cases rose to alarming levels (see figure 24). However, after ten months on the job, the positions of the three epidemiologists were abolished due to a change of government (I later recruited De Quadros and Arnt for WHO service in other countries).

The program staggers to a finish

The work of the three epidemiologists provided a boost to the surveillance effort and dramatically illustrated the need for a sustained effort if Brazil was finally to achieve eradication. However, coincident with the departure

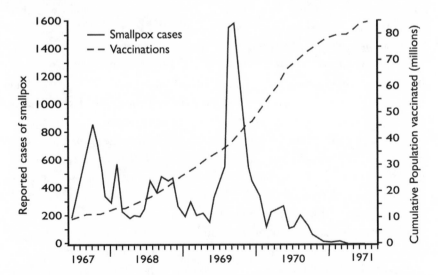

Figure 24. Brazil Reported Cases (1967–1971) and cumulative numbers of vaccinations.

of the epidemiologists, the two key national leaders, Morais and da Silva, also resigned and Morris prepared to leave Brazil to begin a doctoral program. Morris had been the stabilizing force in sustaining the program for three years and in training of staff, despite lack of support from either the regional office or the other three WHO advisers.

Suddenly, the program seemed to be in serious trouble. Help was urgently needed. Working with us in Geneva was a distinguished pediatrician-epidemiologist, Dr. Paul Wehrle, a professor at the University of Southern California, who was on a year's sabbatical. In January 1970 I asked that he travel to Brazil to evaluate the situation, to assess the status of surveillance-containment, and to determine what might be done to sustain the Brazilian program. With reluctant approval from Regional Director Horwitz, Wehrle went to Brazil and traveled extensively. He found that sixteen of the twenty-seven states and territories still had surveillance officers, but reports from a number of key states were greatly delayed and incomplete. He was most surprised, however, to discover that none of the state program staff seemed to know who the WHO regional adviser was and denied ever having met anyone with that responsibility. This was especially puzzling, as Rodrigues had regularly reported extensive, work-related travel throughout Brazil, including the states Wehrle had visited. I had been very disappointed in Rodrigues's performance and several times had proposed to Horwitz that he consider appointing a new regional adviser. With Wehrle's reported observations, I thought he surely would do so. He refused.

Our concern about vaccine quality in Brazil continued—thirty-five of forty-three batches tested at Connaught Laboratories over the preceding three years had failed to meet stability standards. Because of the unusually large proportion of vaccination failures, Wehrle obtained for testing fifteen vaccine vials from field teams and sent them to the Canadian reference laboratory. The results were more grim than we had feared: only two vials met the minimum potency requirements. Horwitz expressed no concern about the deplorable results; he was irate that Wehrle had obtained the specimens without special permission.

Meanwhile, as a result of the sudden increase in reported cases in 1969, the minister of health and the president began expressing profound concern. They sought special assistance from other governments. USAID responded by providing for 30 percent of all program expenditures in 1970 and 70 percent in 1971. A new national director was appointed, Dr.

Claudio do Amaral, who transformed the program. He built a strong surveillance program that eventually became the country's national morbidity reporting system. Additional staff and vehicles were added, and the vaccination program surged ahead: 63 million vaccinations were administered in the 1969–1971 period.

Ironically, the last cases were found not in the Amazon jungle but in the city of Rio de Janeiro in March 1971, less than a mile from the program's national headquarters.

A last regrettable chapter—certifying eradication in South America

We knew it would be difficult for many governments and individuals to accept that a disease so prevalent and so serious as smallpox could ever be eradicated. Special measures to prove its absence in each area and country would be needed if health officials were to be sufficiently confident that they could cease vaccination and international quarantine measures.

In 1973, South America became the first regional area to have had at least two years without known cases of smallpox. On this continent, then, would come our first experience in documenting with certainty that the disease had been laid to rest. I wanted to have a respected, independent group examine the evidence and decide whether they were satisfied that smallpox had been eliminated. Only if the evidence was sufficient and credible would they certify the country as smallpox-free.

There was precedence for a certification process. The malaria program had been engaged in a process to certify certain defined areas as being malaria-free. Based on our growing experience with the epidemiological behavior of smallpox, we decided that two smallpox-free years should pass before an area could be declared smallpox-free. So far, eight months was the longest interval that had elapsed in an endemic country between the occurrence of what appeared to be the last case and the discovery of other cases. Twenty-four months was stipulated in order to provide a margin of safety.

National health authorities from each of the countries in South America were asked to prepare thorough reports, documenting their vaccination programs and the capability of their surveillance structure to detect cases. WHO staff and consultants were to provide help where needed in conducting confirmatory studies and in assembling data. In Brazil, during the two years of continuing surveillance, special WHO smallpox eradication teams worked with colleagues in malaria eradication, conducting a village-by-village search for cases throughout the entire Amazon River Basin. We believed that if undetected smallpox cases were present anywhere, it would most likely be in this area. No cases were found.

I had high hopes for the credibility of an international commission report. The commission was to be composed of knowledgeable individuals who had played no direct role in the program they were examining. We expected them to spend two to three weeks reviewing reports, visiting the areas and countries that they deemed most likely to harbor possible cases of smallpox, and undertaking any other relevant activities they believed necessary. At the conclusion of their work, we requested one of two conclusions: (1) that they were satisfied that smallpox had been eradicated, or (2) if they had doubts, they were to spell out the measures required for them to be fully satisfied.

This first experience with certification was profoundly disappointing. At the insistence of the Brazilian government, PAHO named a former secretary of public health of Brazil as chairman. Two of the four other members had been directly involved in the program. The prepared reports for countries other than Brazil were sparse, and the deliberations were described by one of the commission members as "convivial"—in essence, more social than scientific.

Fortunately for the reputations of the commission, the program, and WHO, no further cases were found in South America. We handled subsequent commissions entirely differently as a result of this final, regrettable performance of a disappointing program.

INDONESIA—A REMARKABLE ACHIEVEMENT WITH FEW RESOURCES

Planning and discussions with Indonesian health officials were needed if a program was to be launched. Such activities had been blocked by SEARO's regional director through most of 1967. But in the autumn, four events happened in rapid succession that launched an eradication initiative. These resulted eventually in Indonesia becoming the first endemic Asian country to eradicate smallpox. First, in September, a new SEARO director, Dr. Herat Gunaratne, was elected and took office in February of the following year. A former director-general of health services in Sri Lanka, he had stated categorically that the eradication of smallpox was wholly achievable and that—unlike his predecessor—he supported the program. Second was the appointment, in November, of a new director-general of communicable diseases in Indonesia, Dr. Julie Sulianti Saroso. A decisive, energetic woman, she was determined to bring new life to a discouraged, poorly paid health staff. Third was a SEARO decision in November to promote a dynamic young epidemiologist, Dr. Jacobus Keja, from the Netherlands, to be regional smallpox adviser—freeing him from service under a sour, indecisive superior who had pledged to eat a tire if India ever became smallpox-free. (We eventually delivered a tire to him, but he declined to fulfill his promise.) Fourth, an epidemic of smallpox broke out in Jakarta, the capital of Indonesia. The outbreak began in August 1967 and by November the government was recording more than fifty cases a week.

In November 1967, Sulianti Saroso requested urgent help in developing a plan for a smallpox eradication program. Keja immediately flew to Jakarta and worked out a strategic plan as well as its projected costs and needs. From Jakarta, he flew to Bangkok to attend our first intercountry

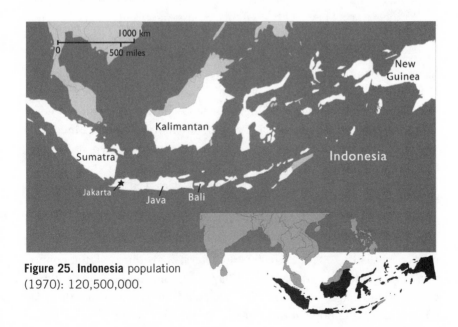

Figure 25. Indonesia population (1970): 120,500,000.

WHO smallpox program meeting. Accompanying him was Indonesia's director of communicable disease control. I met with them immediately after the meeting, and by late December we had a plan and a signed agreement—at a breathtaking pace compared to how the program was progressing in Brazil. We placed orders for vehicles and bicycles, arranged for vaccine contributions, and began recruiting a WHO medical adviser, with hopes of launching the program in June 1968.

I had serious misgivings about undertaking the ambitious plan that had been outlined with the few resources that WHO and the Indonesian government could make available. But there was no alternative other than to push ahead and hope for the best. I had spent months prior to coming to WHO working over detailed budgets for the West Africa program and was reasonably familiar with probable funding needs for national programs. For the West Africa program, we requested and eventually received an average of some $5 million per year in US international assistance. For Indonesia we could make available no more than $250,000 per year from our WHO budget—this to deal with a population that was greater than that of *all* the West Africa countries combined and perhaps as difficult with regard to transportation and communication. A serious fault in the plan, as I saw it, was that it anticipated the effective participation of some 3,000 local vaccinators. They had been in place for some twenty years, but their pay scale had not changed. They received the equivalent of two dollars a month and had to rely on several different jobs to feed themselves, let alone their families. It was suggested that we supplement their salary from WHO funds to permit them to work full time—say, an additional ten dollars a month. This was not possible. It would consume our entire budget and there were other needs—such as vehicles, motorcycles, and bicycles. I was assured, however, that the vaccinators were dedicated people and that, somehow or other, the Indonesians would make the program work. It turned out that the vaccinators worked almost as well as was hoped, but how the vaccinator payment problem was solved is still unknown.

Indonesia posed a number of unique geographical challenges. Its population of 120 million was scattered over 3,000 islands in an area spread over more than 3,700 miles (6,000 kilometers). Two-thirds of the population were residents of the islands of Java and Bali, and more than 80 percent lived in villages of fewer than 5,000 people. It was estimated that only 60 percent of the population could be reached by car or motorcycle. For

20 percent, a boat was required, and another 20 percent could be reached only on foot or by bicycle.

During the 1930s, in the Dutch colonial period, a comparatively primitive but effective vaccine facility had been developed that produced a dried vaccine that was heat-stable, although heavily contaminated with bacteria. Nevertheless, using this vaccine, the country became smallpox-free in 1937 and remained so until 1947, when smallpox was reintroduced. The vaccine-production facility was still in use in 1967, but the equipment had deteriorated and the quality of the vaccine was poor. Accordingly, vaccine production was shut down for a year and a half, the facility was renovated, and new equipment was provided by UNICEF. Dr. Colin Kaplan of the Lister Institute was instrumental in revamping the production methodology; meanwhile, interim donations of vaccine were received from New Zealand, Thailand, the USSR, and the United States. By late 1969, the laboratory was producing enough good-quality vaccine to meet Indonesia's needs.

Dr. Petrus Koswara, the Indonesian director of the program (see figure 26), and the WHO staff member, Dr. Reinhard Lindner from Austria, were

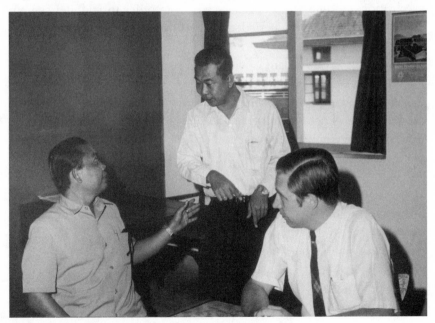

Figure 26. Dr. Petrus Koswara (*center*), national program director for Indonesia, with the author on the right and a provincial medical officer. *Photograph courtesy of WHO.*

eager to get the program under way. But the obstacles were serious: months-long delays in getting the vehicles and bicycles through customs; recurrent governmental fiscal crises, which meant that payments to staff were periodically interrupted and vaccine production activities suspended; and problems in communicating with the many scattered islands across the archipelago.

The program began in July 1968 in Java, one of the world's most densely populated areas. Thirteen "advance teams" were established. Each had a vehicle and was headed by an Indonesian medical officer who had been given a month's special training. WHO supplemented the medical officer salaries so they could serve full time. The teams' primary functions were to work with local authorities to promote the vaccination program, to provide some sort of supervision to vaccinators, and to try to establish the regular reporting of cases to the national smallpox program headquarters. At first these activities seemed to have little effect on smallpox incidence. However, as the importance of surveillance-containment became apparent, the teams in two of Java's three provinces cut back on all routine vaccination and undertook special searches in order to find and contain cases. It was during this period that the first use was made of school-children and teachers to report outbreaks, and the idea of the WHO Smallpox Recognition Card came into being (see plates 2 and 3). As the number of smallpox cases declined to low levels in a province, the teams moved on to more heavily afflicted areas. It was a surprise to find that despite the density of population, the spread of smallpox remained concentrated geographically. Long-distance spread to more distant areas was infrequent.

In Indonesia, the reported numbers of cases of smallpox for 1967 and 1968—13,000 and 17,000, respectively—depicted a problem that was much less serious than the program staff had anticipated. Keja, then serving as the SEARO regional adviser for smallpox, decided in 1968 to undertake a population-wide survey of facial smallpox scars. From this he could develop an estimate of the actual numbers of cases that had occurred during 1967. He found that the true number was more likely to have been at least 100,000 cases. The minister of health was profoundly skeptical and asked that his own Indonesian statisticians review the data and reach their own conclusions. They concluded that the true number of cases was actually even higher, more likely 200,000 to 500,000 cases.

THE WHO SMALLPOX CLINICAL RECOGNITION CARD

The Smallpox Clinical Recognition Card (see plates 2 and 3) came to be widely used in all smallpox programs. On one side, it showed a head-and-shoulders view of a young child on the eighth day of rash. On the reverse side were two full-length pictures of a patient, lying on his back and stomach, to illustrate the distribution of the pocks—more concentrated over the face and extremities. Close-up pictures of the hands showed pocks on the palms and soles of the feet.

This card was an innovation inspired by an Indonesian. Among the many workers engaged in 1969 in a door-to-door search for cases was one who regularly reported exceptional success in discovering infected villages. This was puzzling, because he was considered to be one of the laziest of the searchers—one of the first to return from the field each day. When asked the reason for his success, he admitted that he didn't go from house to house; instead he visited only the schools. There he showed children and teachers pictures of smallpox cases, from a WHO teaching folder on smallpox that had been prepared for Africa. The Indonesian children had little difficulty in recognizing the pustules of smallpox—and, as we soon discovered, children between seven and twelve years of age seemed to know everything that was going on in all of the villages.

Cards were made up with pictures of Asian children and were used by searchers—in schools, at markets, and in house-by-house searches. Sealed in a plastic container, they were virtually indestructible.

Interest in the program at the highest levels of government soared, and additional Indonesian government resources were quickly made available.

The development of a surveillance system was one of the more remarkable achievements in Indonesia. It became fully effective in early 1970. The architect was an Indonesian medical officer, Dr. A. Karyadi. He standardized reporting forms and established a goal of receiving all reports within two weeks from provinces and within three weeks from the outer islands. This required imagination and innovation. The postal service was limited in its geographic scope and was unreliable at best. But creative methods were found—enlisting bus drivers, military personnel, special messengers, and businessmen as couriers. By September, Karyadi reported that 95 percent of the weekly reports were being received from all reporting sites on two of the main islands. This contrasted to the situation only a year before, in which only half of the units had reported—with delays of twenty-one weeks. In May 1970 he began issuing a weekly surveillance report, much as Leo Morris had done in Brazil.

As the advance teams experienced increasing success, routine vaccination efforts declined, case searches increased in number and intensity, and additional vaccinators were enlisted to help in containment vaccination. What appeared to be the last cases in Indonesia were discovered in December 1971—little more than three years after the program had begun (see figure 27). Four weeks passed without cases, and then forty-five cases were notified from a subdistrict only seventeen miles (twenty-eight kilometers) from the capital. Special teams began a rapid search and vaccination effort, but 160 cases were discovered before the last occurred in late January 1972.

The success of the Indonesian program, given the obstacles and paucity of resources, was a remarkable achievement. International support was minimal, amounting to only $1.3 million (little more than one cent per person); it included 24 vehicles, 430 motorcycles, and 3,100 bicycles. Several of the exceptional senior staff were eventually recruited to serve as WHO advisory staff in other countries. One of them, the Indonesian program director, Koswara, forty-three years of age, was the only WHO staff

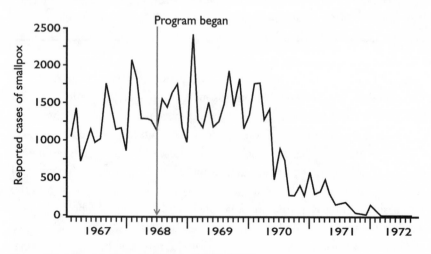

Figure 27. Indonesia Reported Cases (1967–1972).

WANTED: FOR REPORTING SMALLPOX CASES, A REWARD IS OFFERED

In many countries, after the number of cases had decreased to a very few, national authorities and WHO began offering a financial award to encourage the reporting of possible smallpox cases. This proved to be extremely useful—especially in areas where health officers might be inclined to suppress reporting.

The use of a reward for reporting a case was instigated in Indonesia. It was conceived after the discovery, in December 1971, of forty-five cases and six deaths in a subdistrict that had recorded its last cases months before. This was puzzling: five months earlier, a village-by-village search of the area had been conducted, but no cases had been found. Investigators learned that cases had been first reported to the medical officer of health by the subdistrict's vaccinator as early as December 1970. The officer periodically organized ineffectual mass-vaccination campaigns but suppressed the reports of smallpox, fearing that he would be punished for not having achieved a sufficiently high vaccination coverage. Indonesian authorities were concerned that there might be other outbreaks and decided to offer a transistor radio to anyone reporting an active case of smallpox. Numerous suspected cases and illnesses of all types were reported, but no further outbreaks were found. However, the offer of a reward had proved useful in obtaining reports of suspect cases, and a precedent was established.

---

member to die while working in the program. He succumbed to a heart attack in 1974 in Ethiopia.

CERTIFICATION OF ERADICATION—A SERIOUS EFFORT IS MADE

In April 1974 Indonesia became the second country to be eligible for certification of eradication. I was determined that the certification process would be done well—unlike what had happened in Brazil—and that a pattern be established as a model for other countries and areas. Otherwise, there was no possibility of ever persuading the world community that global eradication had been achieved. I had not attended the South American Commission meetings, nor had I reviewed the documentation beforehand, and this clearly had been a mistake. Therefore I became personally involved in the Indonesian certification planning.

As it turned out, the performance of the Indonesian government and staff, the members of the commission, and the WHO advisers was exemplary. Certification was a totally different exercise from that conducted in South America, and this established a pattern of excellence for others to follow.

There were seven people on the commission, including three senior health officials whose countries had special concerns about Indonesia's true status. They were from Malaysia, Singapore, and Australia. The Australian, Dr. Noel Bennett, was highly skeptical of Indonesia's success and expected to find the worst. He stated during the initial meetings that he knew for certain that there were cases of smallpox in Jakarta, and that he expected to discover them when the commission visited the field. The chairman was Dr. Paul Wehrle, who earlier had been a consultant for the Brazilian program. He had visited Indonesia one year before the commission meeting to review what was being done and to suggest additional steps to be taken.

Two WHO staff, Giuseppi Cuboni, from Italy, and William Emmett, from the United States, who had spent the preceding four years in Indonesia, worked with national staff to prepare the documentation. Data from all provinces provided detailed confirmatory information including the regularity of weekly reports; vaccinations performed; numbers of rumored cases investigated (8,505); specimens processed for smallpox virus (1,758); the results of searches in all villages with reported smallpox between 1970 and 1972 (1,352); and special surveys in the more remote areas of two island provinces (27,538 children in twenty-two locations).

At the end of a two-day introduction to the program, Dr. Sulianti Saroso said to the commission: "Go wherever you wish, ask anyone whatever you desire to know. We are convinced we have had no smallpox for more than two years, and we want you to feel as confident as we do." The commission members each went to different areas and spent a busy two weeks in the field. Bennett spent his entire time in the poorest socioeconomic areas of Jakarta. On return, they unanimously agreed that Indonesia had eradicated smallpox. Bennett, however, worried as to what he would tell his Australian colleagues: "My mates will never believe me." He had been an ideal choice for the commission!

In addition to setting a standard for certification, the program itself and the Indonesian staff had previously made another important contribution at an interregional seminar, held in New Delhi, India, in December 1970. By that time, it was becoming clear that after less than three years, Indonesia was well on its way to stopping smallpox transmission. The staff offered to present a number of papers documenting, in particular, the successes they were having with the surveillance-containment strategy. India, at that point, had been engaged in its massive smallpox eradication effort for more than seven years—and did not appear to be much closer to eradication than when the program began. Many Indian participants doubted the Indonesian reports, emphasizing in particular that the Indian health system was vastly superior to that in Indonesia. The results, however, spoke for themselves, and less than a year later Indonesia was dealing with its last outbreak. The Indonesian success turned out to be an important stimulus to India and to other programs in Southeast Asia, which were struggling with similar problems of densely populated areas and a long history of traditional local vaccinators.

Chapter 5

AFRICA

A Formidable and Complicated Challenge

It was difficult to know how to tackle Africa. Almost every country either had endemic smallpox or was the immediate neighbor of one that did. The population of Africa was 364 million spread across an area equivalent to more than three times that of China or the United States. It was politically complex as well, with forty-five independent countries and four large colonial territories—Angola, Mozambique, Namibia, and Southern Rhodesia. There were difficult problems of every sort: civil strife, famine, hordes of displaced refugees, devastating infectious diseases, a severe shortage of health resources and staff, poor roads and communications, authoritarian and often unstable governments—and, in some countries, little support for a smallpox eradication program, let alone any number of other disease-control initiatives. Eradicating smallpox here would require cooperation from every country on the continent and differently tailored eradication plans for each. No program had ever endeavored to accomplish a task as ambitious and overwhelming as this.

Most African countries were members of WHO and were served by the African Regional Office (AFRO), located in Brazzaville in the Republic of Congo. That office was its own challenge—difficult to reach by air and hard to contact by any type of communication. Its small staff was expected to deal with the daunting health problems of all independent countries on the continent, except for the North African countries and Sudan, Somalia, Ethiopia, and Djibouti (until 1977 named the French Territory of the Afars and Issas). At the same time, the office had a deserved reputation for being one of the most poorly managed and uncooperative of WHO's six regional offices. It was clear that we would not receive much help from this source. In fact, the office caused many more problems than it solved.

With the array of impediments we faced, I never dreamed that in 1972, just five years after the eradication effort began, we would be recording the last cases of smallpox on the entire continent, with the exception of two countries—Ethiopia, whose program was then scarcely a year old and Botswana, just beginning to cope with smallpox imported from South Africa.

Considering Africa as a whole in 1967, it seemed reasonable to accept government reports that the 71 million people living in the five countries of northern Africa (Egypt, Libya, Tunisia, Algeria, and Morocco) were free of smallpox and probably would not need special assistance from WHO. All five reported that they had stopped endemic smallpox before 1961, and although most shared common southern borders with endemic countries, the formidable barrier of the Sahara Desert shielded them from imported cases.

I hoped that the four areas that were under colonial authority and South Africa would be able to deal with smallpox without WHO's intervention; this hope proved correct. South Africa did require special encouragement and an official visit before it took action, but it reported no smallpox after 1971. (Programs in the three countries in the eastern horn of Africa—Ethiopia, Somalia, and Djibouti—are described in chapter 9.)

The numerous African national programs proved so distinctive in their challenges and methods of operation, and so diverse in character, that I can only highlight some of the quandaries they presented and how the teams dealt with them. (Each is described in detail in other publications, including *CDC and the Smallpox Crusade* and *Smallpox and Its Eradication.*)

The first programs to begin were those supported by the United States in West Africa with advisory staff provided by the US Centers for Disease Control (CDC). The program covered a population of 116 million people in some of the world's poorest and most heavily infected countries. A second major problem was Africa's two largest countries geographically, Zaire, then known as the Democratic Republic of the Congo (21 million) and Sudan (14

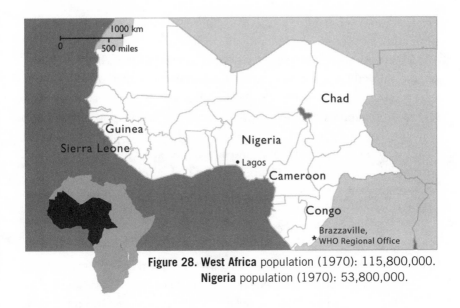

Figure 28. West Africa population (1970): 115,800,000.
Nigeria population (1970): 53,800,000.

million). A third region consisted of the seven countries of eastern Africa (50 million), and the fourth comprised the two countries and four colonies of southern Africa (45 million) (see figure 28).

AN EARLY START IN WEST AFRICA

The decision by the United States in late 1965 to support the eighteen-country West Africa program (later increased to twenty countries) for smallpox eradication and measles control had come at a fortuitous moment. This initiative almost certainly tipped the balance in the 1966 World Health Assembly in favor of undertaking the intensified global eradication program. Moreover, the program provided crucial insights into the epidemiology of smallpox in Africa. The epidemiologists soon discovered that smallpox spread more slowly and was less contagious than medical texts suggested. Early detection of an outbreak and vaccination of a few hundred people in the immediate vicinity was usually enough to stop an outbreak. It was far easier than we had expected. Moreover, few cases occurred in people who had *ever* been successfully vaccinated. Thus, vaccination campaigns could focus on reaching unvaccinated children because most adults were protected by previous infection or vaccination.

The initiative for the West Africa program had begun when I was chief of the surveillance section at the CDC. As described previously, we drafted a proposal to the US Agency for International Development (USAID) to support a five-year program for the eradication of smallpox and the control of measles in West Africa. I had not expected that more than a portion of the proposal would actually be accepted. To my stunned surprise, the proposal was approved and funded in full in November 1965. I was then made director of the program, a responsibility I held until October 1966 when I became director of the WHO global program.

USAID authorities had expected that it would take at least three years to begin the West Africa program because of the complex logistics in the many countries involved. Before we could even begin operations, we were required to obtain signed agreements from each of the governments; after which there would inevitably be delays in the recruitment and training of staff and in procurement of supplies.

I felt strongly that we had to begin in less than three years if we were to take full advantage of the interest and momentum of US and national officials. We arbitrarily set a beginning target date of January 1967, with the hope of getting eight to ten programs under way in 1967 and the rest in 1968. Various political demands forced us to increase the target for the starting number to sixteen countries; meanwhile, two additional countries were added to the list to make it a twenty-country program. I confess that had I known how difficult it would be to set in motion this multicountry, complex operation, I would have been inclined to delay the start for another year. However, I had no prior experience in establishing an international initiative of this nature, nor had others at the CDC. And so, quite innocently, we forged ahead.

Amazingly, plans, agreements, resources, and advisory personnel were actually in place for all but two of the sixteen countries by March 1967. We asked WHO to reimburse some local costs that could not be provided by the countries themselves or by the bilateral assistance program. I was told that this was unprecedented. Subsequently, there was much grousing in WHO about gasoline bought with WHO money being used in USAID-donated cars. But eventually WHO relented and provided about $200,000 each year.

We hoped to begin a training program in July, barely six months after the program was authorized. Recruitment was a pressing priority. We planned to post thirty-seven advisory personnel, primarily medical epidemiologists and operations officers in Africa. These staff members would be backed up by a small regional office in Nigeria and an eight-member central support team in Atlanta. For qualifications, we focused primarily on motivation, intelligence, a willingness to live and work under difficult conditions, and youth. Previous experience was not a factor, as only a few American candidates—mostly those from our own CDC smallpox unit—knew much about smallpox as a disease or had worked in a large-scale vaccination program in a developing country. Almost a generation had passed since smallpox had been endemic in the United States, and few vaccination programs abroad were supported by US foreign assistance.

What we needed was a training program that included a crash course in smallpox—a unique "how-to" training program covering operational methodology, epidemiological and clinical characteristics of smallpox, surveillance methods, and administrative logistics. We quickly prepared a two-hundred-page *Manual of Operations*, which we later modified for WHO use in other countries. To get the program under way, several problems had to be overcome. For example, USAID required that all vehicles be US-made. But in all of West Africa at that time, there were no US automobile dealers, parts suppliers, or mechanics. I arranged to have all staff receive a weeklong course in auto mechanics, and made plans to develop auto parts warehouses in each country. Another challenge: the staff would be using jet injectors, one to administer subcutaneous measles injections, and a second to give intradermal smallpox inoculations. Thus, we scheduled another training course—on how to use and repair these injectors, which were fairly high-maintenance. Finally, we set up intensive French language training for families going to French-speaking countries, along with courses in the history and customs of the countries of West Africa. The summer of 1966 was a frantic one.

In October 1966, I left for Geneva to assume direction of the global program, and Dr. Don Millar (see figure 29) took over my job at the CDC. Millar had directed our CDC Smallpox Unit since 1961. He had been eager to obtain public health training in London but deferred doing so until several of our planned smallpox vaccine studies had been completed. However, the 1965–1966 academic year looked like it would be compara-

Figure 29. Dr. Donald Millar, chief of the CDC West Africa Program from November 1966 to 1970. Earlier, as chief of the CDC Smallpox Unit, he supervised the first field studies of the jet injector. *Photograph courtesy of CDC.*

tively uneventful for smallpox, so Millar left for London. In late autumn we became deeply involved in creating the West Africa program. So much for my prescience!

We were fortunate in recruiting an exceptionally talented and experienced administrative officer, Billy Griggs. He was sorely needed. Dealing with supplies and logistics for twenty different national country programs and twenty different customs administrations was a massive job in itself. It was doubly compounded by USAID's legendary bureaucracy.

Regional leadership in Africa would be provided from a base in Lagos, the capital of Nigeria—a country whose inhabitants accounted for nearly half the population of the countries in the entire program area. The team's director was a veteran of other African programs, Dr. George Lythcott, assisted by Dr. Ralph Henderson (no relation to the author). Also based in Lagos was Nigeria's national adviser, Dr. Stanley Foster (see figure 30), who was destined to spend the next fifteen years working in smallpox eradication. The CDC field advisory staff consisted of one or two people for each country and, in Nigeria, one or two for each region. With so few

Figure 30. Dr. Stanley Foster, CDC chief smallpox adviser for Nigeria, seen here using a Ped-o-Jet Injector at a celebratory function. The instrument was used throughout West Africa for both smallpox and measles vaccination. In 1972, Foster became the chief smallpox adviser for Bangladesh. *Photograph courtesy of CDC.*

advisers, it was clear that the African national program staff would have to be the foundation and principal operators of this program. Advisers would serve as catalysts and support for planning and evaluation.

It was necessary to obtain from each country a signed agreement to the plan of operations—an understanding of what the government and the United States would each provide, provisions for duty-free entry of supplies, and operational understandings. The process of obtaining these generally went well. The French-speaking countries were almost all members of regional French-sponsored health organizations and were familiar with the concept of mobile disease-control teams. In fact, six had already participated in US-supported measles vaccination programs in the early 1960s. In those countries the smallpox eradication and measles control activities were integrated into an established and respected effort.

Likewise, agreements with the English-speaking countries were readily established with one exception—Nigeria, the largest country and the one intended for our regional headquarters. Without an agreement, the program was paralyzed. There were two roadblocks, one caused by the government and the second by the US embassy. Nothing could proceed without the approval and signature of the president of Nigeria—but getting access to him seemed to be impossible. At that time, Eastern Nigeria was threatening to secede, so he was understandably preoccupied with matters of state. We asked the US embassy to help. However, USAID's firm policy in Nigeria had been to support only economic development projects. The USAID mission director and the ambassador were openly hostile to the smallpox-measles program and were reluctant to cooperate—no matter what policies had been decided in Washington, DC. They insisted they could do nothing to help us get the president's agreement—nor could the United Nations resident representative.

As a last desperate move, I requested Lythcott to leave the training program in Atlanta and fly to Lagos to see what he could do. Lythcott was a highly skilled and experienced African American physician, extraordinarily gifted in dealing with people. Nevertheless, six weeks went by without apparent progress. I was fearful. The success of the entire program depended on the Nigerian agreement being signed. However, Lythcott steadily worked his way into the highest diplomatic social circles. Finally, at an evening reception he met a young woman who turned out to be a very close personal friend of the Nigerian president. He persuaded her of the importance of the program. The next morning the last of the needed national agreements was signed.

Surveillance-containment is renamed "Eradication-Escalation"

One of our first field operatives in Nigeria was the very tall, irrepressible Dr. William Foege (see figure 31), whom I had recruited from his Lutheran mission post in Eastern Nigeria. Foege had previously worked with me at the CDC, most recently in the smallpox unit, and he welcomed the challenge of the new eradication program. (He was eventually to become director of the CDC and later of the Carter Presidential Center.) In November 1966, before most personnel and equipment had reached the field, Foege and two CDC staff members arrived in Eastern Nigeria. They soon received reports of smallpox cases from missionaries and quickly worked to control them. Using borrowed motorbikes and a limited supply of vaccine, they successfully contained three outbreaks simply by vaccinating household and village contacts. They discovered that the first cases had come into the area from Northern Nigeria. In an effort to discover other outbreaks, they recruited a missionary-based radio network to help report cases; subsequently, other health units were included in the network. The smallpox cases were occurring primarily along the northern tier of the region, and so, as more vehicles and vaccine became available, they focused on this area. By the end of May 1967, they had detected and contained 754 cases and vaccinated about 750,000 of the 12 million Eastern Nigeria inhabitants. Cases had rapidly diminished in number, and finally some weeks went by with no new cases being discovered. On May 30, Eastern Nigeria proclaimed itself the independent nation of Biafra. Active fighting broke out, and the CDC team was forced to flee the country. Later, Red Cross workers and others working in Biafra reported that they had encountered no smallpox cases. Smallpox transmission appeared to have been stopped by a vaccination program that had reached less than 10 percent of the population.

Foege concluded that the surveillance-containment component of the eradication strategy could prove to be more effective than we had anticipated—that it might be possible to stop transmission even before a mass-vaccination campaign could be completed. A year later, in May 1968, he proposed this to the CDC staff for implementation throughout West Africa. The effort was to be labeled "Eradication-Escalation." The one country where implementation of the strategy was delayed was, ironically, Nigeria. There had been serious concern that the Biafran civil war might spread more widely throughout that country and stop activities altogether. Thus, it was decided to put all possible effort into vaccinating as many

Figure 31. Dr. William Foege, the CDC's smallpox adviser in Eastern Nigeria; he later became a member of the Atlanta headquarters unit. He joined the SEARO Central Advisory Team in India in 1973. *Photograph courtesy of CDC.*

people as possible as quickly as possible, in order to limit the size of potential subsequent outbreaks should the program be interrupted by war.

Meanwhile, the mass-vaccination programs using the jet injectors had proved to be remarkably effective. Special teams checking on vaccination coverage found that vaccinators were reaching more than 80 percent of the villagers. By September 1968, when the "Eradication-Escalation" strategy got under way, almost 60 million of the targeted 110 million vaccinations had already been given; fifteen of the twenty countries were free of smallpox. In Nigeria, with nearly half of the West Africa population, smallpox incidence had fallen dramatically (see figure 32). The special strategy did play a significant role, however, in Guinea and Sierra Leone, where operations began a year after the other programs.

The last cases in the whole of West Africa were thought to have occurred in October 1969, less than three years after the program had begun. Until March 1970 surveillance and search operations failed to detect other cases. An evening to celebrate the success of the program was in progress when a report was received of a suspect smallpox case

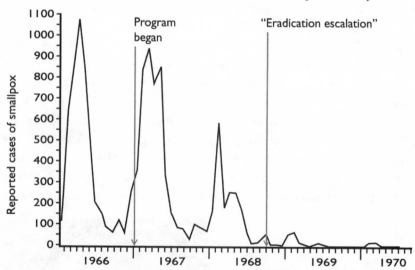

Figure 32. Nigeria Reported Cases (1966–1970). By the time the surveillance-containment (eradication-escalation) program began, mass vaccination had largely contained smallpox transmission in most of West Africa, including Nigeria, the most populous of all. A strengthened surveillance program, however, was essential in certifying eradication.

admitted to a hospital in Northern Nigeria. Foster himself drove some two hundred miles to the hospital, confirmed the diagnosis, and undertook a nighttime emergency vaccination program. Some thirty active and recovering patients came through the line. In all, seventy-five cases were eventually discovered before the outbreak was stopped. The final case occurred in May 1970—the last in West Africa.

The fact that this bloc of countries became smallpox-free in such a short time provided a considerable boost in confidence that eradication was feasible and that the overall strategy of mass vaccination coupled with surveillance-containment worked well. The total American expenditures over five years amounted to $31 million; of this, 40 percent was for the measles vaccination component. The amount originally budgeted had been $47 million. It was one of the rare government programs to be completed ahead of schedule and under budget.

SMALLPOX IS ERADICATED FROM AFRICA'S TWO LARGEST COUNTRIES

Sudan and Zaire covered an area equivalent to half the size of the United States, with about 35 million people distributed widely over the two coun-

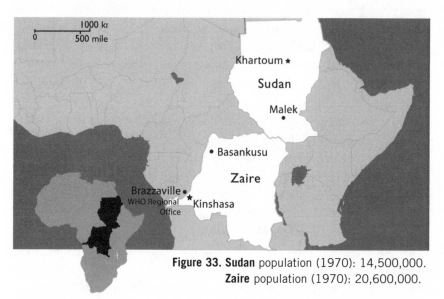

Figure 33. Sudan population (1970): 14,500,000. **Zaire** population (1970): 20,600,000.

tries (see figure 33). They shared a common border but otherwise were a study in contrasts—beginning with the fact that in 1967 smallpox was widely prevalent throughout Zaire but totally absent from Sudan.

Zaire—major epidemic center at the heart of Africa

From the beginning, I was aware that Zaire would present major difficulties. Most of the country is covered by dense tropical rain forest with no more than 1,900 miles (3,000 kilometers) of all-weather roads. Communications, even between the larger towns, were meager to almost nonexistent. Zaire's 21 million people spoke at least two hundred different languages and dialects.

Zaire had become independent in 1960—but within a week, United Nations forces had to be brought in to quell a secessionist movement. Until 1969 a United Nations technical assistance program provided overall administrative direction; WHO operations were directly overseen from Geneva during this period. There were few educated local personnel: primary education had been first offered by the government only in the 1950s, but few people had attended secondary school or a university. The government was in shambles, and lack of security was a continuing threat in many areas. In the entire nation, there were only ninety physicians.

As I learned more about these problems and what they meant to field operations, I became increasingly convinced that of all the places on earth, this was the one where we were least likely to succeed in stopping smallpox. But we had to achieve it: one-third of all cases in central, eastern, and southern Africa were being reported from Zaire (and this despite a reporting system that was the most incomplete of any country). Zaire shared common borders with five countries, all of which were smallpox-free or experiencing only a few cases. Zaire was to receive more than 8 percent of all WHO funds each year through 1974.

A plan for the operational strategy had been drawn up in 1966 before I had arrived in Geneva. Basically, it was a traditional blueprint, calling for vaccination teams to work systematically throughout the country. Afterward, some sort of maintenance program would be implemented. Nothing was said about surveillance. However, the plan was unusual in that it also called for the inoculation of BCG (tuberculosis vaccine) at the same time as smallpox vaccine. BCG was provided by UNICEF. Although BCG was only marginally effective, its use was widely advocated. The BCG program had begun a year before. The WHO adviser responsible for it was given overall responsibility for the smallpox program as well. The addition of BCG complicated operations enormously because BCG had to be administered very superficially (intracutaneously) by syringe and needle. It required two people to hold the child's arms firmly so he would not move while a vaccinator inserted the needle.

A pilot program began in 1967, but the numbers vaccinated were far fewer than had been planned. One reason was the decision by the adviser to give a special certificate with an official stamp to each vaccinee and to enumerate by household the exact number vaccinated. Several months after the program began, jet injectors were brought in to speed up the pace of smallpox vaccination—but for each jet injector, six clerks were needed just to fill out the certificates.

By September 1968 it was clear that the program was in trouble. New leadership was needed. The Geneva-based United Nations international assistance office for Zaire appointed as codirectors of the program Dr. Pierre Ziegler and a young, highly motivated Zaire physician named Lekie Botee (see figure 34). Ziegler, a French medical officer, had directed mobile prevention and treatment programs in Chad for sixteen years. During 1967 he had worked with CDC advisers in Chad in implementing the smallpox eradication-measles control program. He knew well the practicalities of running field operations. Among their first official directives was the order to stop the time-consuming distribution of certificates and the individual registration of households. The pace of vaccination began to pick up.

The difficulties were not over, however. In 1969 overall United Nations responsibilities for Zaire were transferred from Geneva to AFRO. Although the AFRO office was located in Brazzaville, directly across the river from Kinshasa, Zaire's capital, communications between the program office and AFRO were negligible and assistance was nil. Two years went by before the promised WHO advisory staff positions were finally filled. Meanwhile, I suggested the possibility of using jet injectors to administer the BCG vaccine. Studies had shown that the results with the jet injector were comparable to those of BCG given with syringe and needle. This change was immediately vetoed by AFRO.

A serious, recurrent problem was the government's failure to release funds for the program for weeks or even months at a time. As a result, all operations would periodically grind to a halt as funds ran out. An inventive

Figure 34. Drs. Lekie Botee and Pierre Ziegler. Botee was the first Zairean codirector of the smallpox eradication program and later became director-general of Zaire's Health Services. **Zeigler,** French, WHO codirector of the program, was a veteran of sixteen years of African mobile health programs in Chad. *Photograph courtesy of WHO, 1968/G. Prethus, 1970.*

solution was urgently required. Ziegler, with the help of a part-time finance officer and veteran of Zaire service, Seve Axell, from Sweden, persuaded the government authorities to deposit allotted funds for the program in separate bank accounts to which only he and Ziegler had access. When one principal account ran out, as it regularly did, Ziegler would inform the ministers of health and finance and issue dire warnings of the implications of having all activities stop. Nevertheless, weeks would pass before funds would again begin to flow. Meanwhile, the program mysteriously carried on without interruption, thanks to the auxiliary bank account.

The challenges faced in this vast country were beyond those encountered anywhere else in Africa but were not beyond the skills of the remarkably creative Ziegler and Botee. The roads were in terrible repair; vehicles broke down so often that it became necessary to create an all-purpose vehicle repair and maintenance workshop (see figure 35). The bridges were an even greater hazard. They had been constructed by the Belgians prior to 1960 but had not been maintained; some had been destroyed during the civil war and hastily repaired for temporary use in transiting rivers. National staff for the program had to be trained as supervisors and managers due to the low levels of formal education. Sustaining as few as three vaccination teams was difficult, especially given the paucity of WHO advisory staff. The reporting of smallpox cases remained grossly incomplete, but there were not enough personnel to establish a surveillance-containment program.

Not until early 1971 was the full WHO staff of eight finally in place, augmented by an eight-man US Peace Corps contingent. One of the staff was an especially dedicated former volunteer, Mark Szczeniowski, who was to remain in Zaire for seventeen years playing a key role in investigations of monkeypox subsequent to eradication. For communication directly with the Kinshasa headquarters, 100-watt transceivers were installed in trailers that accompanied the vaccination teams. Remarkably, the target of 24 million vaccinations was reached by July 1971 (see figure 36). Five surveillance-containment teams had begun work in January; with the end of vaccination, this number increased to eleven. Smallpox cases fell sharply during 1970 and 1971; in June 1971 the last case was detected—less than three years after Ziegler and Botee assumed direction.

I found it difficult to believe that the country was really free of

Figure 35. Smallpox Vehicles—Zaire. The Zaire road system had not been maintained in years and was devastated by years of protracted fighting. Broken chassis were common, and a special repair and maintenance workshop was essential. *Photograph courtesy of WHO.*

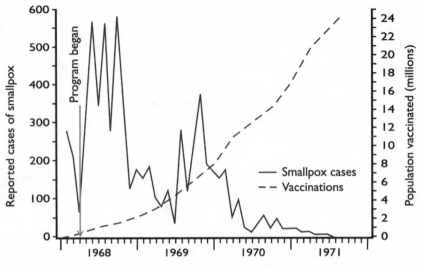

Figure 36. Zaire Reported Cases and Cumulative Vaccinations (1968–1971).

smallpox. But an intensive search during the succeeding years detected only sporadic cases of human monkeypox. Countries around Zaire became free, and none detected importations.

As in West Africa, it seems likely that the most significant changes in the smallpox control strategy included the appointment of strong, creative leaders such as Ziegler and Botee and the provision of ample supplies of freeze-dried smallpox vaccine. Surveillance programs soon were able to confirm that transmission had indeed been interrupted. In these sparsely populated areas where transportation was difficult, people did not travel widely and smallpox transmission was not readily sustained.

Sudan—a smallpox-free country becomes infected

When the global program began, Sudan was considered to be free of endemic smallpox. It had become smallpox-free in 1963 during the course of a mass-vaccination program using freeze-dried vaccine—and except for two small imported outbreaks had remained so until 1968. Although Sudan shared a long common eastern border with highly endemic areas of Ethiopia, there was confidence nationally and in WHO that Sudan could readily deal with any imported cases. Its health system was far better developed than those of either Ethiopia or Zaire.

THE FIRST CASE OF MONKEYPOX—A CHALLENGE IN ZAIRE

On September 1, 1970, a nine-month-old child was admitted to a clinic hospital in the small town of Basankusu in central Zaire. He had a rash that looked like smallpox and had enlarged cervical lymph nodes like those of a patient with mumps. The physician was puzzled. He decided to obtain a specimen for examination at the WHO reference laboratory in Moscow. Dr. Svetlana Marennikova, the director of the laboratory, eventually reported that she had isolated the virus of monkeypox from the specimen—a virus never previously isolated from a human case.

The isolation of monkeypox virus raised a serious question. Clinically, the case had appeared to be smallpox. Was it possible that there might be an animal reservoir of this virus that would jeopardize the eradication program? Dr. Ivan Ladnyi, the AFRO regional smallpox adviser, flew down from Nairobi to join Ziegler in a search for other possible cases and to evaluate the status of vaccination and surveillance in the area. It was not an easy task. The 62,000 residents of the Basankusu Territory were mainly primitive farmers living in small villages scattered along paths and tracks in the dense tropical rain forest. One all-weather road crossed the 180-mile-wide territory, but it was intersected by several impassable rivers with neither bridges nor regular ferry service. Travel along the tracks and paths was perilously difficult during the ten-month rainy season.

There were three options for getting to Basankusu from Kinshasa: (1) a weekly air service that was often canceled; (2) travel by road, which meant crossing no fewer than twenty lakes or rivers, only some having a regular ferry service; (3) a twice-monthly river boat to the provincial capital where, after a wait of several days, another boat made a two-day trip to Basankusu. The team elected to fly.

Eighteen government and mission dispensaries served the area. All were now reporting regularly; none had seen cases of smallpox in more than two years. Ladnyi and Ziegler did a scar survey in the area from which the patient came. Of 1,132 people examined, 94 percent had vaccination scars. No other cases could be found. However, many of the villagers regularly ate monkeys and small rodents. As we later learned, rodents were the probable source of infection.

This was but the first of many cases of monkeypox to be reported over the months and years ahead.

Extensive deserts stretched across northern Sudan, giving way to steppe and grasslands and then to marshes in the tropical south. The large irrigated agricultural areas in the central part of the country drew more than 700,000 migrant workers each year from the southern provinces of Sudan and Ethiopia. Imported cases from Ethiopia were first detected in 1968. They increased sharply in 1969. Over the next two years migrant laborers spread smallpox across the country.

In 1967 Sudan had requested help in starting a mass-vaccination program, asking that a WHO adviser be assigned. The program began in January 1969, but it was poorly conceived and ineffective. As in Zaire, the national authorities, on the misguided advice of the WHO adviser, decided to give BCG vaccine along with the smallpox vaccine. As might be expected, this made operations much more complicated. National staff believed that the Sudanese people would not gather at collecting points to receive the smallpox vaccine as they did in West Africa. Thus, the plan was for smallpox vaccination teams to go house to house. This was impractical for BCG vaccination, which required multidose syringes. Thus, those up to age twenty were asked to assemble at collecting points to receive BCG. Extensive written records were compiled—a list showing the name of the head of each household, the number and age of residents, and the number vaccinated. It was the same approach that had proved to be disastrous in Zaire. Two complete sets of records for each village were dutifully and laboriously compiled, but they were never compared to each other—nor, for that matter, were they used for assessment or follow-up vaccination. There was no provision for a surveillance system. Five million vaccinations a year were targeted, but the actual numbers fell far short of the target.

In 1970 and 1971, more than 1,000 cases of smallpox were reported each year, but this was likely not more than 5 percent of the actual totals. The campaign was failing. Smallpox had reestablished itself throughout most of the country. The cases were *Variola minor*, which was milder than the *Variola major* found in Zaire and West Africa, and this made it easier for the disease to spread. Even if sick, patients could still travel. The smallpox regional adviser, Dr. Ehsan Shafa, an Iranian, tried to convince the program's directors to abandon temporarily the labor-intensive use of BCG and to emphasize surveillance. The WHO national adviser refused to permit this. In August, cholera broke out, and so smallpox vaccination ceased altogether as the teams began a mass cholera vaccination program.

THROUGH THE BACK DOOR—GETTING VACCINE THROUGH A WAR ZONE

In 1970, I was visited in Geneva by representatives of the Anyanya Resistance Movement—rebel forces who were active in southern Sudan. They said that smallpox was not present in their area, but they wanted the vaccine because they were afraid of smallpox being brought into the area by Ethiopian migrants. When we asked how we could get vaccine to them, they said that they regularly took supplies from northern Uganda into Sudan, traveling on foot for seven to ten days through the forests. It was in everyone's best interest that they have the vaccine—but the Sudanese government could not vaccinate in the rebel-controlled areas, and WHO could not provide vaccine to the Anyanya. The dilemma was resolved by giving quantities of vaccine directly to the resistance movement's leaders and recording the amount as "lost to inventory." After the war ended in 1972, Sudanese staff found that there were surprisingly large numbers of vaccinated residents in the southern rural areas, but no cases.

Cholera vaccine was recognized to be even less effective than BCG, but it was popular with politicians. It was seen as a demonstration that the government was doing something about the epidemic.

In late 1971—more in desperation than with hope—Shafa and I decided to convene a special seminar in Khartoum, the capital. We brought together physicians and senior sanitary inspectors from each of the provinces of Sudan to discuss surveillance and containment and to describe for them how successful this approach had been in West Africa, Brazil, and Indonesia. According to one of the senior Sudanese attendees, Dr. Omar Sulieman (see figure 37), the seminar was a turning point for him. He could envisage this strategy being especially effective in Sudan. He requested of the government that he be given direction of the national program. WHO offered to assign a second adviser, but, unhappily, the WHO adviser turned out to be even more incompetent than the first one. Sulieman alone transformed the program.

Sulieman was a tall, formidable figure with a booming voice who radiated confidence and authority. He was at least six feet tall and wore a great turban, which added significantly to his stature. He directed all health

Figure 37. Dr. Omar Sulieman established an effective program throughout Sudan in January 1972, and within one year, transmission ceased. In 1973, he became an adviser in Pakistan and revived a flagging program. *Photograph courtesy of D. P. Francis.*

workers and the police to report suspected smallpox cases. Special health staff investigated and contained smallpox outbreaks as rapidly as they were reported. A peace treaty was signed with the Anyanya resistance in March, and Sulieman quickly moved to recruit their military personnel to search distant areas for new cases. About fifty medical students joined the program as well.

On December 17, 1972, the last outbreak of smallpox was detected and contained in a village called, ironically, "Malek"—the local name for smallpox. This happened just six months after workers detected Zaire's last case. The surveillance-containment component of the strategy was crucial to success in Sudan.

EASTERN AFRICA—A MASS-VACCINATION ACHIEVEMENT

Programs in the five former English colonies (Kenya, Malawi, Tanzania, Uganda, and Zambia) and the two former Belgian colonies (Burundi and Rwanda) (see figure 38) proceeded with little direct input from me or our Geneva staff, once they had received modest resources from WHO. In these countries the health services, roads, and communications were generally better developed than in West Africa. Many people were being vaccinated regularly through clinics and special campaigns. However, there were many vaccination failures due to use of locally produced liquid smallpox vaccine or, in Rwanda, a substandard dried vaccine. Still, the countries had managed to control smallpox reasonably well.

When I had arrived in Geneva in late 1966, little was known of the potential for any of the eastern African countries to develop eradication programs and whether any would need WHO assistance. I asked AFRO to help us by contacting national health authorities. The office refused. As a matter of policy, it said, the office responded only to requests for assistance originated by the countries themselves—and none had asked for help in eradicating smallpox.

Breaking the impasse required imagination. Dr. Stephen Falkland, one of my staff, pointed out to AFRO staff that the director-general was

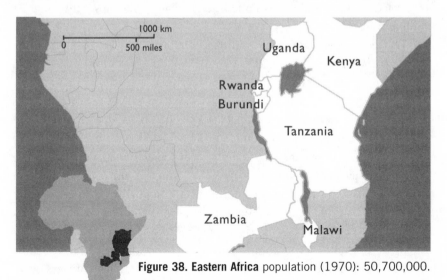

Figure 38. Eastern Africa population (1970): 50,700,000.

obliged by the assembly to make a report on the smallpox program at the May 1967 World Health Assembly. Information regarding these countries was important, so Falkland proposed that he travel to several of the countries to "gather information." The regional office agreed, noting that the countries were not to be encouraged to request WHO help. Falkland took along draft plans of operations and letters of request for the possible signature of the respective ministers of health—just in case they were needed.

Falkland made brief visits to Burundi, Kenya, Rwanda, Tanzania, and Uganda, pointing out to each that if assistance was desired, WHO could provide vehicles, vaccine, needles, an adviser, and money to cover the costs of petrol and vehicle maintenance. Each country would be responsible only for the salaries of national staff, many of whom were already employed by the BCG vaccination program. With the savings on smallpox vaccine and transport, the costs to the governments for the eradication programs would be marginally more than the vaccination control programs they already had in place. Each of the countries "spontaneously" requested assistance.

Due to delays in delivery of vehicles and other supplies, national programs did not get started until 1969. Meanwhile, we began to supply the countries with freeze-dried vaccine and to help upgrade vaccine production in Kenya. By the end of 1968, Kenya was supplying all of its own vaccine. Eventually it supplied neighboring countries with some 15 million doses.

A major impetus for success in these countries was Dr. Ivan Ladnyi, a Ukrainian, who served as AFRO's regional WHO adviser for all of eastern and southern Africa. Ladnyi had been recruited to his post in Kenya in 1965. At first, his work was severely constrained by AFRO, whose office was located in the Republic of Congo, a considerable distance away. The regional office required that he get special permission for each trip before he could travel anywhere. Six to eight weeks usually passed between the time he requested permission to travel and the time he received a response—usually denying him permission. Eventually, Ladnyi found it expedient to simply telex his request for permission two or three weeks ahead and to proceed on his trip expecting to find on his return a telex saying he couldn't go. It was a mystery to me as to why he was never chastised by AFRO for insubordination and why his travel vouchers were always reimbursed—no questions asked. I suspect there were two possible reasons—one was the administrative ineptness at AFRO and the other, the reluctance at that time of WHO staff to question the behavior of Soviet assignees.

Ladnyi was key to catalyzing programs all over eastern and southern Africa as well as Zaire—until 1971, when his assignment to WHO from the Soviet Union was ended by his government. I pleaded with Soviet Vice Minister Venediktov to extend Ladnyi's assignment. He told me that Soviet policy forbade an extension, but he agreed to send him back in one year if I would write a letter commending Ladnyi's performance. I did this only to learn some months later that Ladnyi had been promoted to a position that Venediktov identified as the third-highest ranking health position in government. Venediktov told me later that he had never met Ladnyi before receiving my letter. He decided to summon him for an interview and was so impressed that he offered him a prominent position in Moscow. He pointed out that Ladnyi was too important to the Soviet health department to be assigned back to the smallpox program. (In 1975, however, Ladnyi was named WHO assistant director-general—he became my boss, in fact.)

As mass-vaccination programs progressed in eastern Africa, new cases of smallpox decreased rapidly. In 1971 surveillance-containment teams began to be activated, about the time the mass-vaccination campaigns ceased and just as the last endemic cases occurred. Remarkably, this was just over two years after the programs had begun. Although the surveillance-containment activities played no role in interrupting the transmission of smallpox, they were useful in showing that smallpox really had been eliminated. Later, they proved to be a necessity in Uganda to contain imported cases from Sudan in 1971 and 1972, and in Kenya to contain imported cases from Ethiopia and Somalia through 1976.

SOUTHERN AFRICA

We knew there was smallpox in southern Africa, but gleaning precise information was difficult (see figure 39). There were problems in communication with each of the countries or territories in this group except for Botswana. Angola and Mozambique were Portuguese colonies in 1970, and Portugal was not particularly communicative, nor did it welcome WHO contacting its colonies to obtain data. Angola apparently was smallpox-free, but Mozambique reported cases up to 1970, although none thereafter. The country was embroiled in civil war, and cases of smallpox were occurring regularly in areas bordering Southern Rhodesia (now Zimbabwe). Southern Rhodesia itself was in the process of breaking away from the United Kingdom and had no communications with WHO. South Africa's membership in WHO had been suspended in 1964 because of its apartheid policies. Each of the countries had ongoing vaccination programs. I had wanted to know a great deal more about the status of smallpox in that region. But for the first four years of the program, there was neither time nor resources nor authority for me to investigate.

South Africa reported cases weekly to WHO in compliance with the requirements of WHO International Health Regulations and registered 100 to 200 cases of smallpox each year. No information was forthcoming as to where the cases were occurring or what South Africa was doing to control the disease. *Variola minor* had been the prevalent form of smallpox in South Africa for more than fifty years. As in Brazil, it was regarded as a minor illness, not much more serious than chicken pox.

In 1970 South Africa unexpectedly took a new interest in eradicating smallpox. This interest stemmed from an annual status report on smallpox that I had prepared for the WHO executive board in January of that year.

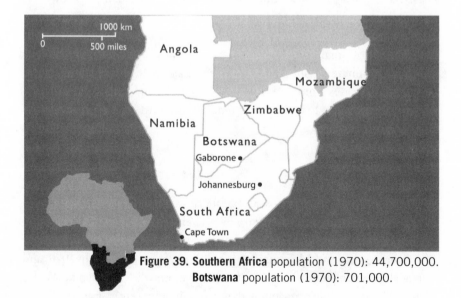

Figure 39. **Southern Africa** population (1970): 44,700,000.
Botswana population (1970): 701,000.

AN UNPRECEDENTED VISIT TO SOUTH AFRICA—JULY 1972

In June 1972 it appeared that the countries of southern Africa might be free of smallpox. No cases had been reported since South Africa's notification of seven cases that January. We needed to know more about smallpox in South Africa in order to gauge the risk to other African countries. I wanted to make a personal visit to assess the situation, but United Nations officials were not permitted to visit South Africa. However, a World Health Assembly resolution had requested the director-general "to continue to take all necessary steps to assure the maximum coordination of national and international efforts." This was interpreted by Director-General Mahler as license for me to visit.

I was cordially received by the South Africans—but as an American, not as a WHO staff member. The official government records for 1972 showed only the seven cases that had already been reported to WHO; all were from a hospital 180 miles from Botswana. However, a visit to the National Institute of Virology revealed a surprise. The laboratory itself had recorded ten poxvirus isolations in 1972, the last occurring in May. None was from the January outbreak and none had been reported to public health officials. I asked for additional information and eventually I learned that there had been more than twenty cases involving patients, staff, and visitors at a large hospital in Transvaal State, adjacent to Botswana. The outbreak had continued until August.

The hospital staff had not been more diligent because of a communications failure. In the South African laboratory, suspect smallpox specimens were inoculated into tissue cell cultures rather than being grown on fertile hens' eggs. In tissue culture, it is impossible to distinguish variola virus from *vaccinia* virus; therefore, the director issued a report stating that the virus discovered was one of the "*vaccinia-variola* group." He believed that it would be clinically obvious to the attending physician whether the case was one of smallpox or vaccinia. The hospital staff, however, did not understand the distinction. They reclassified their patients as having disseminated *vaccinia*, rather than smallpox. In consequence, smallpox transmission had continued and spread to Botswana where more than a thousand cases occurred before it could be stopped.

I wrote: "Of the endemic countries in Africa, South Africa and Ethiopia are the only ones which have not yet initiated eradication programs.... The continuing reservoir of smallpox in South Africa and Ethiopia is of increasing concern to neighboring countries." I pointed out that South Africa was the only country in Africa still using the liquid vaccine, which deteriorated rapidly at normal temperatures. South Africa sent an angry letter to the director-general, claiming that health officials were operating an effective control program and that they didn't need freeze-dried vaccine. However, as I later learned, the South African government, shortly after sending the letter, began to produce a freeze-dried vaccine. In June the country started a mass-vaccination campaign in the northern province of Transvaal, which had reported most of the cases. It was a modest effort, but reported cases decreased rapidly in number; none were reported after January 1971.

The Botswana debacle

By the summer of 1971, only four and a half years after global eradication began, it seemed to me that we might have a smallpox-free Africa perhaps as early as 1973. Only Ethiopia and Sudan were then reporting cases—but programs were in place in both countries.

This did not happen. We discovered that Ethiopia had much more smallpox and more serious widespread problems than we had anticipated. South Africa's last cases were reported in January 1971, but neighboring Botswana found a suspect case five months later, in June, and four more in August. Immediate investigation and control measures were called for, but AFRO repeatedly blocked all offers of assistance. Fully eight months passed before Botswana mounted effective actions to stop transmission. By then, the disease had spread throughout the country. In neighboring countries, vaccination teams and the military went on the alert endeavoring to protect their borders.

The Botswana epidemic had begun inauspiciously. On June 1, 1971, we received a report in Geneva that a single case of smallpox had been isolated at Gaberone Hospital in the capital city. This was followed by a cable on June 7, stating that smallpox had been confirmed virologically by the government laboratory in South Africa. At that time the only smallpox in Africa since January 1971 had been in Ethiopia, a very long way from

Botswana. I telegraphed AFRO to point out the urgency for investigation and containment. A week later, on June 16, I received a cable from AFRO, stating that the situation had been misunderstood and that the patient actually had chicken pox. Given the laboratory findings this made no sense. I sent a return reply urging that an epidemiologist from our office visit immediately. AFRO replied that such a visit was not considered advisable.

From the WHO side, we had not planned to provide special assistance to Botswana, a small, largely desert country with a population of only 701,000. I had not been particularly concerned about the risk of smallpox being imported. Botswana had experienced no smallpox cases since 1964 when it had controlled an imported outbreak by mass vaccination. Since then it had been vaccinating about 40,000 people each year. Eighty percent of the population lived on a strip of land not more than 120 miles wide but the area was adjacent to Transvaal—which, as we learned later, had been South Africa's most heavily infected province.

At the end of August more cases of smallpox were reported. There was another flurry of communications between Brazzaville and my office in Geneva resulting from a September 20 message received from Botswana urgently requesting freeze-dried vaccine and an operations officer to help control the epidemic. Instead, AFRO decided to send a Pakistani medical officer, Dr. Zia Islam, who had recently arrived in Kenya to replace Ladnyi as the regional smallpox adviser. He arrived in Botswana on October 8 for a fifteen-day visit. He was a willing worker but knew little about smallpox or surveillance and had received no briefing from smallpox program staff before going there. This was not surprising, as there was no one in the Brazzaville office to brief him, and regional directors regularly refused to send new appointees through Geneva for orientation. He accomplished nothing.

More cases of smallpox occurred, and we continued to press AFRO to appoint an adviser. As before, the response was "We can take no action unless we receive an official application from the government." Finally, in November, I had a fortuitous discussion with the resident representative of the UN Development Program in Botswana, who was then visiting Geneva. He grasped the seriousness of the situation, and soon after his return to Botswana, the government sent a formal request for an operations officer. In February 1972—eight months after the first case had been reported—Garry Presthus, a US veteran of the Zaire program, was transferred by AFRO to Botswana.

Presthus, with an experienced Botswana senior health inspector, Joseph Sibiya, quickly organized surveillance and mass-vaccination programs. Over the next seven months, 1,105 cases were detected and 500,000 vaccinations administered. In March 1973, just as they thought the last outbreaks had been contained, a case was discovered, followed by nineteen more. Those infected were members of a Christian religious sect of about 4,000 members that traveled in migratory bands throughout southern Africa. They resisted vaccinations of all types. The only feasible control measure was to mandate vaccination—and this the president of Botswana did, bluntly stating that the sect, Mazezuru, could either accept vaccination or be deported. They accepted.

THE LESSONS OF AFRICA

I was elated by the remarkably rapid and successful interruption of smallpox transmission in Africa. Certainly there were innumerable difficult problems—social, political, and geographic—but the innovation and dedication of the national and international staffs had overcome them with surprising facility. It was apparent that competent national program leaders and a few capable international advisers could accomplish wonders. In most of Africa the principal impact factor had proved to be effectively managed mass-vaccination programs using good-quality freeze-dried vaccine. Surveillance-containment was a significant added component in stopping transmission in a few West African countries as well as in Sudan and Botswana. However, it was essential in all countries for confirming eradication and in dealing with importations.

By the summer of 1973 we were left with only five endemic countries —India, Pakistan, Bangladesh, Nepal, and Ethiopia. We were soon to confirm a now familiar picture: the challenges posed by each of them were unique and far more complex than we had ever imagined. We had acquired considerable expertise and knowledge of the behavior of smallpox, but we had a very long, steep, arduous learning curve ahead about smallpox in South Asia that would tax the best we could offer.

Chapter 6

INDIA AND NEPAL

A Natural Home of Endemic Smallpox

From the beginning of the global program, the challenge of endemic smallpox in India loomed like a massive, ominous thundercloud. In 1967, India's 550 million people accounted for more than half of the total population of the endemic countries—and more than half of all reported smallpox cases (see figure 40). Temples to Sitala Mata, the Hindu goddess of smallpox, were present throughout the countryside. Smallpox had been so long in India and so prevalent that many believed that its heavily populated Gangetic River Plain might well be the endemic home of smallpox—from which this disease, like cholera, could never be eradicated. Whatever the case, our entire budget of $2.4 million was much too small to make an appreciable dent in the problems of a country so large or on a disease so entrenched.

For the first few years, most of my time and the WHO resources were focused on the other endemic countries of Asia, Africa, and South America, in an attempt to pare down the geographical extent of smallpox. Eventually the ultimate battle for smallpox eradication would have to be won in India.

It was not as though India had failed to make a major effort. In 1962, the country had begun an intensive mass-vaccination eradication campaign. However, by 1967 there were as many cases being reported as before the program started. The discouraged and frustrated government health staff, close to abandoning the program, reluctantly staggered on.

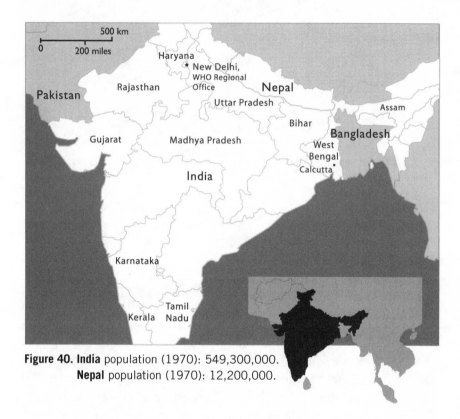

Figure 40. India population (1970): 549,300,000.
Nepal population (1970): 12,200,000.

Nepal, a comparatively small independent country, lay adjacent to India's northern border and epidemiologically was a part of India. Most of the population resided within sixty miles (one hundred kilometers) of the border. An eradication program was not undertaken until 1971, but then it succeeded in rapidly stopping transmission. Subsequently, Nepal had to contend with repeated importations from adjacent, highly infected Indian states.

AN AMBITIOUS NATIONAL PROGRAM IS CREATED—1962

The program in India was one of the first to be launched after the 1959 World Health Assembly's decision to attempt global eradication. This was done on the advice of the Indian Council of Medical Research. The council recommended starting pilot studies immediately to determine methodology and costs, and to begin a full-fledged five-year program in 1962. The plan called for 152 operational units, each with ninety people to vaccinate village by village, with the intention of reaching 80 percent of the population within three years. After this, local primary health centers were to vaccinate the newborns and others who might have been missed. The program also called for a small advance team to create a written record of all residents in a log book—name, address, age, sex, and previous vaccination history. Vaccinators would follow and enter the date that the vaccination was performed. All subsequent vaccinations would be entered in books that were expected to be used for at least twenty years. The record keeping alone was a gargantuan and unrealistic undertaking.

The USSR offered to provide freeze-dried vaccine and eventually supplied 450 million doses. The United Nations Children's Fund (UNICEF) pledged equipment for vaccine production. The United States offered $2 million at the beginning of the program and over a five-year period provided some $23 million in financial aid.

Between 1962 and 1966, the teams vaccinated an impressive number— 440 million people. But as the program was concluding its three-year "attack phase," an assessment by India's National Institute of Infectious Diseases produced disappointing findings—vaccine coverages were lower than anticipated, ranging from 54 to 73 percent in various districts. In many areas, little had been done to vaccinate children born after the mass vaccination. Nothing had been done to promote reporting of cases. With the conclusion of the attack phase, the work of the vaccination teams ceased. For maintenance vaccination, one vaccinator was provided for every 10,000 to 20,000 people. Four freeze-dried vaccine production facilities had been established during the attack phase, but here, too, results were disappointing. As 1966 ended, they were producing scarcely enough vaccine for 20 million people.

THE PROGRAM NEARLY COLLAPSES—1967–1969

Bad news from India was waiting for me when I first arrived from the Communicable Disease Center (CDC) to assume my new WHO position as chief of the Smallpox Eradication Unit. The Indian government was frustrated to the point of seriously contemplating severe cutbacks in the smallpox vaccination initiative. Worse, some advocated that India formally withdraw from participation in the global eradication program. Such a move would have spelled death to the program. I immediately flew to New Delhi in December 1966 to plead with government officials to keep the program going. They listened but made no commitment. Dr. Chandra Mani, director of WHO's Southeast Asia Regional Office (SEARO), offered no encouragement. An Indian himself and former director of health services, he was convinced that eradication was impossible. He saw little point in encouraging the Indian government to persist in the futile effort.

Unexpected events came to the rescue. The budget and provisions for staff under the five-year eradication program expired in April 1967. However, the numbers of smallpox cases began rising dramatically during the early months of that year; in fact, several states were reporting major epidemics. The government was deeply concerned. It agreed to continue the program pending the results of a study by a combined WHO-India team. It was agreed that sixteen people would spend six weeks in the autumn of 1967 to make a firsthand assessment of the problems and to offer recommendations.

The findings of their assessment were devastating. The teams documented a lack of personnel and meaningful direction at every supervisory level of government—federal, state, and district. There seemed to be enough vaccinators, but their productivity was low. Reporting systems were disorganized with perhaps 10 percent of cases being reported. Freeze-dried vaccine was improperly stored, and quantities of the heat-sensitive liquid vaccine were still being used. The elaborately prepared family registries were frequently incomplete and generally useless.

With 85,000 recorded cases in 1967—the highest total in a decade—the Indian government decided that there was no choice but to continue the effort. In January 1968, it issued a formal statement pledging to maintain its eradication efforts for an indefinite period. A new director for the national program was named, Dr. Mahendra Singh. He was the only professional in the national office; he had no vehicle and was permitted to travel only by train or bus. No one worked more tirelessly than he in urging the states to improve their programs, but one man with limited authority could do only so much. A two-person WHO regional infectious diseases team, Drs. Jacobus Keja from the Netherlands and Andrzej Oles from Poland, offered in vain to provide assistance. The government insisted that national staff, now with more than five years' experience, could learn nothing from WHO advisers. In fact, neither of the advisers was permitted to travel in India outside of New Delhi without filing a specific itinerary and receiving advance written permission from the government.

I was anxious to know how the surveillance-containment strategy might work in India if it consisted only of prompt weekly reports from all health units and the containment of reported cases by special teams. An early opportunity to test this approach arose in the southern state of Tamil Nadu. At a meeting in 1967, I encountered Dr. A. Ramachandra Rao, director of the Infectious Diseases Hospital in Madras, the capital city. Rao was well-known for his studies of clinical smallpox. He expressed interest in the surveillance-containment strategy and the possibility of organizing a mobile team to promote case detection and reporting, and to investigate and contain outbreaks in the state. I provided WHO funds to cover the costs; he recruited a team that began work in January 1968. The hope that one team could have an impact in a state with 40 million people seemed unlikely, but only 263 cases had been reported statewide in 1967. Tamil Nadu had a comparatively good health service and high levels of vaccination coverage. Between January and June 1968 the team detected and contained fourteen outbreaks, half of which were confirmed to have been brought in from neighboring states. From July 1968 to June 1969 only two imported outbreaks were found in the entire state. Rao's team extended its work to a neighboring state. As he observed, the containment of outbreaks worked well because smallpox was spreading surprisingly slowly. This was as clear a demonstration of the value of the surveillance-containment component as I could have foreseen. We subsequently made Rao's findings widely known throughout India, but they were generally dismissed as being relevant only to a wealthy, comparatively well-vaccinated state like Tamil Nadu.

National freeze-dried vaccine production had increased steadily in quantity and quality but remained insufficient to meet India's needs. As a result, the liquid vaccine continued to be used in a number of areas. Using the bifurcated needles would have been a help because it would permit the standard vial of freeze-dried vaccine to vaccinate one hundred people instead of twenty-five. In mid-1968 we began distributing the needles to programs in all endemic countries. Most countries put them in use as soon as they were delivered—but not India. Several Indian administrative and scientific leaders speculated that the needles might not be as well accepted by the public as the traditional rotary lancets (see chapter 3). They doubted that they would produce vaccination "takes" as frequently. Moreover, they argued that it would be both difficult and time-consuming to retrain the tens of thousands of vaccinators who were then using rotary lancets. Thus, before the needles could be used, a number of special studies had to be conducted to address each concern. Not until late 1969 were all the hurdles finally cleared and the needles put to use. Even so, the uncooperative, heavily populated states of Bihar and Uttar Pradesh remained holdouts, refusing to use the new needles until 1971. Not until 1973 were all rotary lancets replaced—and this happened only after supervisors were ordered to confiscate them.

A RESURRECTION OF THE PROGRAM—1970

Through 1969, reported smallpox cases gradually decreased. This primarily reflected progress in India's more prosperous southern states. However, it was becoming clear that with the sad state of surveillance, the number of reported cases was not an accurate measure of progress. A good reporting system was critical. Another opportunity to demonstrate the surveillance-containment strategy arose in the western state of Gujarat. It began experiencing serious epidemics in 1969 and soon was responsible for one-third of all cases in India—one-fifth of all cases in the world. The state director for the program, Dr. G. J. Ambwani, asked for help, and in April 1970, I traveled to Gujarat with Oles and Singh. Vaccination scar surveys revealed the population to be surprisingly well vaccinated. The cases were primarily occurring in slum areas among migrants who probably made up no more than 10 percent of the population. I related Rao's observations in Tamil Nadu and proposed that Ambwani create two special surveillance-containment teams, operating from his state office, to contain outbreaks. Ambwani acted quickly to do so, and fourteen months later smallpox transmission in Gujarat had been stopped.

That surveillance and containment were as important as mass vaccination remained a difficult concept for program directors and health ministers to grasp. All they had ever known was mass vaccination. Therefore, in November 1970 we convened an interregional seminar in New Delhi devoted explicitly to surveillance and containment. The event brought together smallpox staff from countries throughout Africa and Asia to relate their own personal experiences. India seemed to me to be the best venue, because the meeting would attract many national staff in addition to those traveling from other countries. I asked that all papers be reproduced and distributed widely.

Meanwhile, it appeared from the available data that smallpox in India was moving steadily north—almost like a tidal wave—into the states of Rajasthan and Haryana. These states bordered on New Delhi, the capital. In reporting to the director-general of health services, Dr. J. B. Srivastav, I took pains to note the threat to the capital. I reiterated the recommendation of the 1967 joint WHO-India team—that state smallpox program teams be created to improve reporting mechanisms, as well as to conduct surveillance and containment of outbreaks. I proposed that WHO assign four advisers to India and pay the costs for travel for additional surveillance personnel employed full-time in smallpox units at the national and state levels. Until this point, India had rejected all proposals to supplement Indian staff with WHO epidemiologists. But now, reluctantly, they accepted, and a surveillance program began to take shape. In 1970, for the first time, WHO support for the program exceeded $100,000. During the next seven years, that amount would rise to more than $11 million.

Toward the end of 1970, I was informed by the very capable Keja that, for family reasons, he had to assume another post. Fortunately, I was able to replace him with someone who became a legend in the program—Dr. Nicole Grasset (see figure 41), a Swiss-French virologist and epidemiologist. Grasset assumed the position of SEARO's principal smallpox adviser and held this post through the achievement and certification of eradication in India and Nepal. She had joined the program after working as a Red Cross adviser, providing vaccination and medical care in the Nigerian civil war zone of Biafra. An energetic, determined, charismatic leader, Grasset was the only woman in the entire regional office, except for the nursing adviser. She was quite unlike anyone I have ever met. Her energy and optimism were contagious. She regularly traveled into the field, hiking across fields to distant villages, always impeccably dressed, ever optimistic and energetic. With these attributes and her open dedication to the program, she seemed to be able to relate quickly to everyone from the prime minister to industrial leaders, district officers, and ordinary vaccinators. She

Figure 41. Drs. Nicole Grasset and Zdeno Jezek. Grasset, a Swiss-French medical microbiologist-epidemiologist, was the senior smallpox adviser for SEARO from 1971 through the end of the program. **Jezek,** a Czech physician-epidemiologist, became a SEARO senior adviser in 1972 and later joined the WHO headquarters smallpox staff. *Photographs courtesy of N. Grasset/WHO.*

seemed to know everyone and they knew her. And yet she was a modest person, embarrassed by any sort of personal special recognition.

Grasset was joined a year later by the tireless, self-reliant Dr. Zdeno Jezek (see figure 41), a Czech epidemiologist. Jezek had just completed a two-year assignment with WHO in support of health programs in Outer Mongolia. It was said that if he were to be parachuted into an unexplored area, he would have a map drawn in an hour, the natives organized in two hours, and a health system in a day or two. He worked tirelessly seven days a week. Jezek eventually became part of the Geneva headquarters staff.

AN UNEXPECTED CATASTROPHE IN WEST BENGAL— REFUGEES FROM EAST PAKISTAN—1971

The populous West Bengal State, with its densely crowded capital of Calcutta, had made excellent progress and in 1970 recorded only 172 cases—

Plate 1. Dr. Edward Jenner (1749–1823). English physician and scientist who demonstrated that cowpox could protect against smallpox. The virus became known as *vaccinia* and the process was called vaccination. Pastel by J. R. Smith in 1800. *Photo courtesy of Wellcome Library, London.*

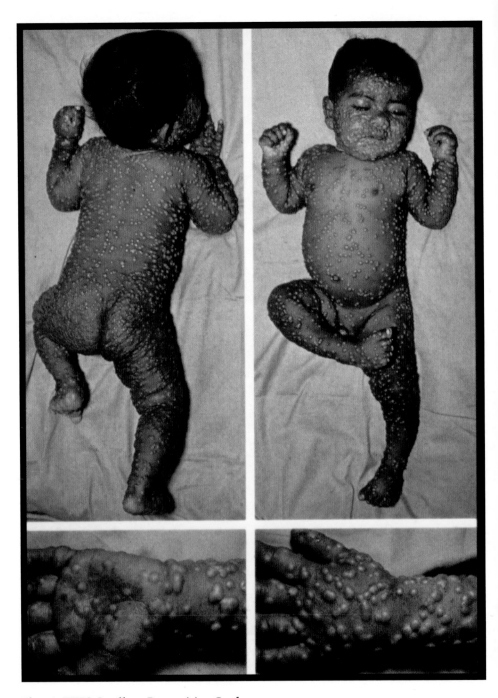

Plate 2. WHO Smallpox Recognition Card.

Plates 2 and 3 are of young children on about the eighth day of the rash. The rash is most dense over the face and extremities. All pustules are at the same stage of development. Pustules are present on the palms of the hands. The pictures are from the WHO recognition card, which was first used in 1972 by search workers. They showed the pictures to villagers and asked if they had seen anyone with such a rash. Neither of the cases shown are severe. *Photo courtesy of WHO.*

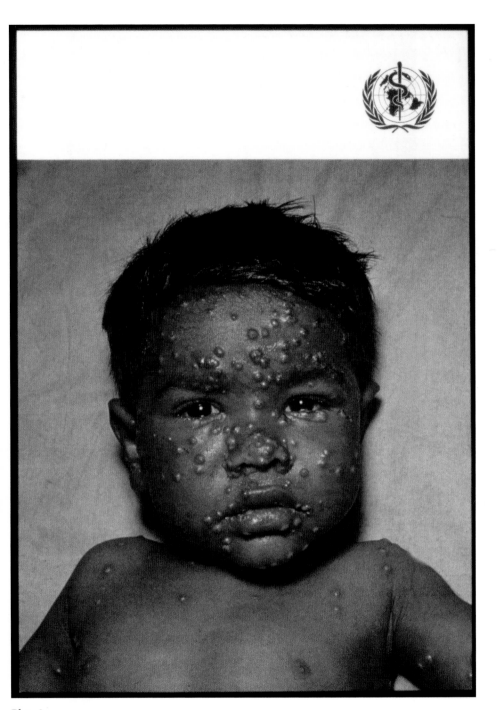

Plate 3.

Plate 4A.

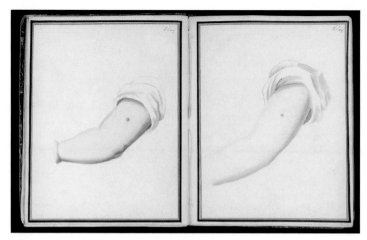

Plate 4B.

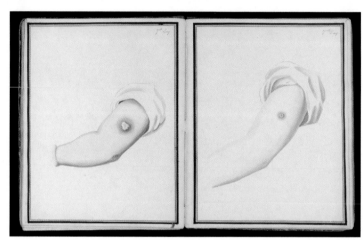

Plate 4C.

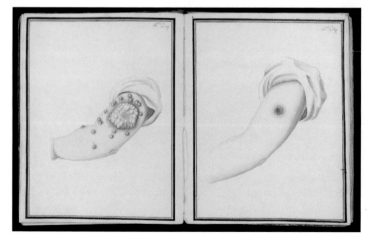

Vaccination vs. Variolation. Watercolor drawings (circa 1802) of local lesions created by variolation (*left*) and vaccination (*right*) on days three, seven, and fourteen. *Photos courtesy of Wellcome Library, London.*

Plate 5. Ali Maow Maalin. The last case of naturally occurring smallpox in the world. He developed a rash on October 26, 1977, in the town of Merca, Somalia. *Photo courtesy of WHO/J. Wickett.*

Plate 6. Smallpox Reward Poster: 1978–79. The WHO poster publicized the reward of US $1,000 for finding a confirmed case of smallpox. *Photo courtesy of WHO.*

نحن أعضاء، اللجنة العالمية للإشهاد الرسمي باستئصال
الجدري نشهد بأنه قد تم إستئصال الجدري من العالم .

WE, THE MEMBERS OF THE GLOBAL COMMISSION FOR THE
CERTIFICATION OF SMALLPOX ERADICATION, CERTIFY
THAT SMALLPOX HAS BEEN ERADICATED FROM THE WORLD.

NOUS, MEMBRES DE LA
COMMISSION MONDIALE
POUR LA CERTIFICATION
DE L'ERADICATION DE
LA VARIOLE, CERTIFIONS
QUE L'ÉRADICATION DE
LA VARIOLE A ÉTÉ RÉA-
LISÉE DANS LE MONDE
ENTIER

我们，全球扑灭天花证实委员会委员，
证实扑灭天花已经在全世界实现。

МЫ, ЧЛЕНЫ
ГЛОБАЛЬНОЙ
КОМИССИИ ПО
СЕРТИФИКАЦИИ
ЛИКВИДАЦИИ ОСПЫ,
НАСТОЯЩИМ
ПОДТВЕРЖДАЕМ, ЧТО
ОСПЫ В МИРЕ БОЛЬШЕ
НЕТ.

NOSOTROS, MIEMBROS DE LA COMISION MUNDIAL PARA LA CERTI-
FICACION DE LA ERRADICACION DE LA VIRUELA, CERTIFICAMOS
QUE LA VIRUELA HA SIDO ERRADICADA EN TODO EL MUNDO.

Genève le 9 décembre 1979

Plate 7. Certificate of Eradication. Official certificate signed on December 8, 1979, by the members of the Global Commission certifying the global eradication of smallpox. *Photo courtesy of WHO.*

Plate 8. *World Health* **Magazine Cover, May 1980**. The issue of the WHO magazine *World Health* was devoted to smallpox at the time of the formal declaration that the eradication had been achieved. *Photo courtesy of WHO.*

its lowest total ever. However, West Bengal was soon to become one of the more challenging problems in India.

In March 1971, civil war broke out in neighboring East Pakistan (Bangladesh), a country that had recently interrupted smallpox transmission. An estimated 10 million refugees fled to India where many were housed in refugee camps. WHO and Indian advisers checked many of the camps to be certain that none had smallpox and that all people were being vaccinated, with one exception—the largest, called Salt Lake Camp, near Calcutta. The camp was the size of a city, with about 250,000 refugees. Responsibility for medical care at the camp had been assigned to a private international relief organization. For unknown reasons, West Bengal authorities refused to permit federal smallpox staff to visit. Smallpox appeared to have been brought into the camp in November but went undiagnosed for nearly two months. The cases were considered to be chicken pox. No smallpox vaccinations were being offered. Not until December 19, 1971, were smallpox cases recognized. An epidemiologist in the United States, watching a television news report, saw what he thought were smallpox cases. He called the CDC, which called me, and I called New Delhi. One of the national program team flew to Calcutta and confirmed the worst: the outbreak was extensive. Vaccination was begun, but it was too late. Bangladesh independence had been declared three days before; 50,000 refugees had already left the camp to return home, and others were on their way (see figure 42). Epidemic smallpox spread across West Bengal as well as throughout Bangladesh. The state reported 4,800 cases in 1972—a twenty-fold increase over the previous year.

THE "FINAL PHASE"—TARGET ZERO—
DELUSIONAL OPTIMISM—1972

Despite the setback in West Bengal, I was optimistic. By the summer of 1972, it was apparent that we would be dealing with only five endemic countries during 1973: India, Pakistan, Bangladesh, and Nepal in Asia and Ethiopia in Africa (see figure 43). For the first time, some sort of surveillance-containment program was in place in each of these countries. The programs needed to be greatly strengthened, but I believed we had a sufficient appreciation of the dimensions of the smallpox problem to

Figure 42. Refugees Returning from the Salt Lake Refugee Camp. In December 1971, Bangladesh became independent and the refugees returned home from India; hundreds of them were infected with smallpox. *Photograph courtesy of UNHCR.*

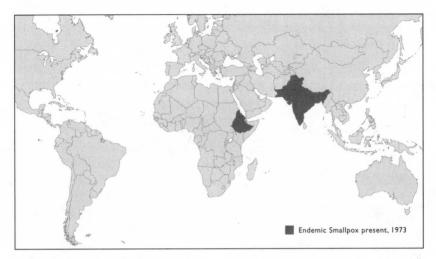

Figure 43. Endemic Countries—1973.

achieve eradication in these countries as rapidly as we had in Africa. I could not have been more wrong.

In September 1973, at a conference in Addis Abeba, I announced the start of an intensified effort called "The Final Phase." The core of the strategy was a greatly heightened emphasis on the surveillance-containment component of the program. I proposed that the goal should be to achieve worldwide eradication by February 1975—a target just eighteen months away. That autumn we also convened special seminars in New Delhi, Karachi, and Khartoum.

In India, I drew special attention to Indonesia, whose program had made such dramatic progress utilizing the surveillance-containment strategy. Although Indonesia had recorded more than 10,000 cases in 1970, transmission had now stopped. The Indian officials expressed doubts about the Indonesian data. They reasoned that if India could not interrupt smallpox transmission, Indonesia—with a much inferior health system—could not do so either. The Indonesians were equally adamant that their data were correct and that their effective weapon had been the surveillance-containment program. The Indian participants were unimpressed.

Shortly after the seminars, we began publishing a monthly report called *Target Zero,* for all senior staff. The publication's name was a pointed reminder that success was not measured by the number of vaccinations but

Figure 44. "Snoopy Smallpox Target Zero" was drawn personally for me by Charles Schulz at a dinner we attended. He wanted me to know that Snoopy supported smallpox eradication even though Snoopy had never been vaccinated himself. *PEANUTS © United Feature Syndicate, Inc. Used with permission.*

by the number of cases of smallpox—and the ultimate target was zero cases (see figure 44).

In India the epidemic tidal wave of smallpox continued to move to the north and east. With the hope of stifling the epidemic before it could get established, I assigned three of the new WHO advisers to states just ahead of the front. In theory, this idea was sound and in one of the states, Rajasthan, the strategy was effective. However, it was soon apparent that in Uttar Pradesh and Bihar, the challenge was far more formidable and the initial efforts were futile. Each state, in population, was larger than any *country* in Africa. They were among the most densely populated and least prosperous of the Indian states and among the most poorly managed.

Special training programs were provided for district health officers, but staff were transferred so frequently to other duties or districts that they seldom had an opportunity to apply what they had learned. And the problems went deeper than mere issues of training. Health officers who reported substantial numbers of smallpox cases were not infrequently punished by being transferred to hardship posts on the grounds that there would be no cases if they had conducted the vaccination program properly. Corruption and bribery were rampant. Progress reports so frequently contained false information that they were nearly meaningless. In Bihar the health workers went on strike, and in Uttar Pradesh the entire staff was diverted for many weeks to administer cholera vaccine. This was known to be ineffective but for political reasons it was given nonetheless.

The heart of the smallpox problem was the three central states—Uttar Pradesh, Bihar, and West Bengal (see figure 45). Their total population was 189 million. Smallpox gained momentum in late 1972 and the winter of 1973. Total cases in India in 1972 were over 27,000—a 50 percent increase over the year before. It seemed likely that the total for 1973 would exceed 60,000. Even where containment programs were already in place, outbreaks were not detected in time to prevent further spread. Surveillance-

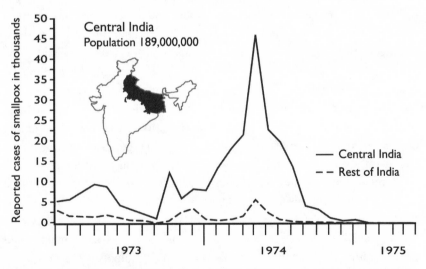

Figure 45. India Reported Smallpox Cases (1973–1975). From 1972, the majority of cases in India were in the three central states of Bihar, Uttar Pradesh, and West Bengal, which accounted for 35 percent of India's population.

containment methods as applied in Africa, Indonesia, and South America seemed to be much less effective. Part of the problem was the density of population; part was an especially mobile population able to move more readily over the extensive network of roads and railways. Often, entire families moved to the more endemic cities. If a member developed smallpox, the whole family then moved back to its rural village and a new outbreak was born.

At the May 1973 WHO World Health Assembly, India's program was pointedly criticized in a manner that was unusual in assembly debates. Delegates raised specific questions as to whether the government was taking advantage of WHO's offer to send emergency aid on request. Director-General of Health Services Dr. J. B. Srivastav was acutely embarrassed and returned to India determined to strengthen the program.

THE ULTIMATE STRATEGY—1973

Stronger leadership and new directions were necessary. At the end of June, I convened a meeting of our WHO staff and senior national Indian health officials to decide how best to meet the challenge. Of first importance was the critical shortage of responsible leaders and supervisors. There was a veritable army of vaccinators and lower-level staff, but until recently there had been only one professional at the national level in India and three WHO advisers. Second, outbreaks needed to be detected far earlier and contained more effectively. Third was the need to improve reporting systems to ensure that cases discovered in the field were accurately and promptly reported through the often dysfunctional chain of government offices before being reported nationally. With "Target Zero" as our objective, it was essential to have an up-to-date assessment of progress.

One of the most important decisions, as it turned out, was to establish a joint Central Appraisal Team. The designated senior Indian official was Dr. Mudi Inder Dev Sharma, director of the National Institute of Communicable Diseases—internationally renowned for his work in malaria and greatly admired by a generation of young epidemiologists (see figure 46). He was soon promoted to commissioner of rural health, reporting personally to the director of health services. Six other highly regarded national public health officers completed the team from India (see sidebar below).

Figure 46. Three Key Figures in the India Program: Mr. Jarl Tranaeus, Dr. Nicole Grasset, and Dr. Mudi Inder Dev Sharma. Grasset and **Sharma** were the principal leaders of the WHO-India Central Appraisal Team. **Tranaeus** was head of the development cooperation office of the Swedish Embassy in New Delhi and was instrumental in obtaining critically needed funds for the India and Bangladesh programs. *Photograph courtesy WHO.*

The WHO national smallpox program advisers, Grasset and Jezek, were joined by Dr. William Foege, who had been a leader in the West Africa program and was sent on loan from the CDC. The fourth of the group was Dr. Lawrence Brilliant (see figure 47), a young American counter-culture physician who had been living on an ashram in India from where he volunteered his services. He had no special qualifications other than that he spoke Hindi and was highly committed. His contributions and those of others of the Central Appraisal Team were catalytic to the success of the program. Brilliant continued in a career in public health and eventually became a senior executive of google.org, the philanthropic arm of Google, Inc.

Individuals of the Central Appraisal Team were expected to travel extensively to problem areas for purposes of oversight, evaluation, and staff motivation. Specific plans were developed for the different states. The high-incidence central states would be given highest priority. Their total popu-

Figure 47. Dr. Lawrence Brilliant, US physician-epidemiologist who joined the Central Appraisal Team in 1973. He is here showing a small version of the smallpox recognition card and asking about the location of other cases. *Photograph courtesy of WHO.*

lation was about 35 percent of India's population. Three southern states—Karnataka, Tamil Nadu, and Kerala—with a combined population of 100 million people—were classified as smallpox-free and would receive little attention. The others, considered "low-incidence" states, would require active programs to eliminate the remaining endemic areas of smallpox and to contain importations. The program created twenty-six special field teams, half of whom were directed by epidemiologists from Indian institutes and universities and half by epidemiologists recruited by WHO. Eventually, a total of 230 WHO epidemiologists from thirty-one countries would head teams for periods of three months to two years. Each worked, along with district health officers, in zones, each with about 10 million people, handling a variety of tasks—training district and local staffs, organizing searches, and developing the surveillance-containment activities.

It was our belief that for the surveillance and containment strategy to work, outbreaks would need to be more promptly detected and containment measures taken soon after. How to do this? In the spring, two pilot studies of significance were undertaken by Dr. Vladimir Zikmund, a

---

KEY LEADERSHIP FOR THE FINAL PHASE

The Central Appraisal Team played a critical role in India in the intensified "final phase" of the program. Using as many as 150,000 workers for special searches and a rapidly evolving set of tactics, they led the effort that stopped smallpox transmission in just twenty-one months.

Indian Government

Dr. Mudi Inder Dev Sharma: director of the National Institute of Communicable Diseases (NICD); Dr. Rabinder Nath Basu: assistant director-general of health services; Dr. Mahendra Dutta: director-general for cholera; Dr. Mahendra Singh: deputy assistant director-general for health services (smallpox); Dr. Sachida Nanda Ray: deputy assistant director-general for health services.

WHO

Dr. Nicole Grasset: SEARO principal adviser; Dr. Zdeno Jezek: WHO staff; Dr. William Foege: consultant, on loan from the CDC; Dr. Lawrence Brilliant: SEARO staff.

Czech WHO adviser, and by my deputy, Dr. Isao Arita. In two different districts they demonstrated that it was possible for health staff, during a one-week period, to visit all villages in a district and for containment vaccination teams to follow up and vaccinate some fifty households around each case detected.

At a June strategy meeting of senior WHO and Indian staff, we decided to attempt a nationwide search for cases, such as Arita and Zikmund had conducted, with priority given to the three states of Uttar Pradesh, Bihar, and West Bengal. There the searches would be conducted in October, November, and December. They would require the participation of at least 35,000 workers. During the first searches, contacts were made with village leaders, schoolteachers, market vendors, and residents in different geographical areas of each village, using the specially designed WHO Smallpox Recognition Cards (see figure 48). As the searches were repeated, they were gradually transformed into house-to-house searches

Figure 48. Schoolchildren and the Smallpox Recognition Card.
Schoolchildren are being shown the WHO smallpox recognition card and asked if they know of cases. Cases within a 10 kilometer (6.2 mile) radius were usually known to children who were between eight and twelve years of age. *Photography courtesy WHO.*

that were validated by independent teams to make sure that 90 percent of all houses had been reached. In the states of lower priority, one or two searches would be done during the October–December period with about 32,000 workers participating. Outbreaks would be contained by local staff with supervision by the special teams. The logistics were staggering, requiring eight tons of forms, schedules, and recognition cards for each search. Nothing comparable had ever been attempted. It clearly was an audacious step to attempt to set a program this extensive in motion within the very short period available. However, epidemic smallpox was spreading.

The searches started in October, the lowest point in the smallpox seasonal curve, with the hope that outbreaks could be stopped before the disease began to spread widely. The rationale was that this should result in fewer chains of infections and lower numbers of cases during the high-transmission period from February to June. We planned to reconvene in January to decide the strategy for the following six months.

Reports from the first search were alarming, even unbelievable. This was the low point in the season—yet in Uttar Pradesh, which had been reporting only 300 cases each week, the first search turned up nearly 6,000 unreported cases in 1,500 different villages. In Bihar, workers found almost 4,000 unreported cases in 600 villages, and even this did not tell the whole story. As we learned later, as many as half of the villages were missed during the first search. There was far more smallpox than we had ever imagined.

The next two searches, fortunately, revealed fewer new outbreaks. Meanwhile, West Bengal looked better than had been expected, reporting fewer than one hundred infected villages at year's end. Only a few hundred cases were found in all of the other states combined. But nearly 90,000 cases were reported in 1973—India's highest total in fifteen years. We assured national authorities that this was primarily a function of better, more accurate reporting—and so we believed.

In January 1974 as strategy was being plotted for the forthcoming months, we were optimistic. Reporting was more complete than ever, and the search programs were steadily improving. It was evident that the basic

CENTRAL APPRAISAL TEAM: TOUGH, COMMITTED LEADERSHIP

The commitment and determination of the Central Appraisal Team staff was as extraordinary as its accomplishments. At the January 1, 1974, strategy meeting of the team, it was apparent to me that they were all exhausted and some were near the point of collapse. All had been working seven-day weeks for four months. They had repeatedly made difficult trips throughout some of the country's most remote and inhospitable areas in a frantic effort to motivate the army of health workers to contain outbreaks. Four of the members had serious medical problems. One had incapacitating renal colic; one had painful facial herpes; another had a serious fungus infection of the foot, which eventually required surgery; and one had atypical pneumonia with high fever and pleuritic pain. However, the only problem they would discuss was where to find the additional resources to keep the program going. I expressed skepticism about their own ability to keep up with the grueling schedule even if resources could be found. Bill Foege replied simply: "We've considered the question, and have decided that things can't get worse; therefore they must get better."

strategy—systematic search, surveillance and containment—needed to continue. In fact, we speculated that it might be possible to interrupt smallpox transmission before the July summer monsoons. There were two problems, however: the funds were all but depleted, and the Central Appraisal Team was exhausted.

More money was required. Within SEARO some funds could be diverted from the Indonesian allotment, but this was far from what was needed. The Americas had just been certified as smallpox-free, but Dr. Abraham Horwitz, PAHO's regional director, refused to permit the funds allotted to his region to be diverted. Repeated requests to WHO member governments brought some vaccine but little in the way of cash. I flew back to Geneva and pleaded the case with Director-General Halfdan Mahler (see figure 49). As it happened, the executive board was to meet in just a few days to decide how to distribute funds that had been provisionally set

Figure 49. Dr. Halfdan Mahler. Danish physician and tuberculosis expert, Mahler was WHO director-general from 1973 to 1988. *Photograph courtesy of WHO.*

aside for health programs in China—funds that China had declined to accept. Mahler immediately agreed that a cable be sent to India indicating that this money could now be used for the smallpox program. The executive board endorsed this diversion of funds after the fact.

This was but the first of several bold decisions that Mahler would make to facilitate the work of the smallpox eradication program. He had only just assumed the position of director-general from Dr. Marcelino Candau, who had served that role for twenty years. Candau had never been persuaded that eradication could be achieved and acted accordingly. Mahler brought the opposite, welcome view.

THE DARKEST DAYS OF ALL—JANUARY TO JUNE 1974

Bihar presented the most daunting challenge by far, and the situation there kept deteriorating. A state of 62 million people, it was infamous for its corruption and criminality. We gradually found that the politically well-connected state smallpox program director had two staffs: a modest-sized official support staff and a second, very large group of clerks and book-keepers, sequestered in extensive quarters unknown to the Central Appraisal Team. He had been given funds to hire several thousand additional vaccinators. Exactly where they were working, however, was a mystery. It was eventually learned that they showed up only once a month—to sign a pay slip and to collect about one-third of their allotted pay. The director pocketed the rest. It took months before the state minister finally agreed to transfer the program director to another position. In a way, I felt sorry for the man, as he had thirteen daughters and needed somehow to provide dowries for each if they were to get married.

Meanwhile, Dutta and Foege of the Central Appraisal Team, working with district staff, struggled to cope with a growing epidemic. Any optimism we had felt in January vanished with the results of the February search. It revealed 1,170 new outbreaks—twice as many as in December. Smallpox was spreading rapidly to the east and into West Bengal. Many more districts became infected (see figure 50). In Bihar reported cases grew rapidly in number, reaching 36,000 in May. Compounding the problem was a strike by Indian airlines, making it impossible to ship vaccine or for the Central Appraisal Team to move quickly between the capital and the states. Next, the railway workers went on strike. At about this same time, the global oil crisis began to be reflected in India as gasoline costs doubled and inflation spread. Finally, the health workers threatened to go on strike. Drought in southern Bihar—serious enough to require international assistance—was followed by monsoons in the north and the most severe floods in a decade. Hundreds of thousands of displaced persons were moving from place to place. Smallpox moved with them. I have often been asked, at what point of the program was I in greatest doubt of success for India and for the program? My feelings of despair during the first six months of 1974 are etched in memory.

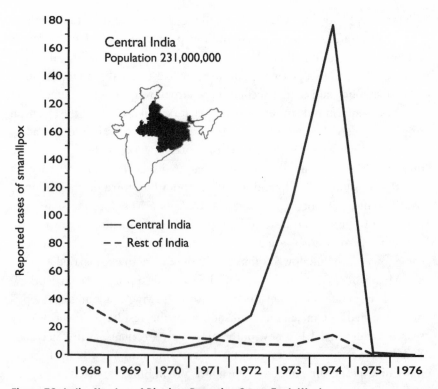

Figure 50. India: Number of Districts Reporting Cases Each Week.
In India during this period, there were 393 districts, each with an average population of 1,500,000 people.

Yet another catastrophe

In late April, formerly smallpox-free areas in India began to report smallpox outbreaks—primarily among laborers. These workers had come from a major railway center and industrial complex, Jamshedpur (population 800,000), in southern Bihar. By early May, 125 such notifications had been received, and ten to fifteen new ones were arriving daily. Brilliant arrived to find a totally disorganized health system and a poorly vaccinated, heavily infected population. Jamshedpur was the major industrial center for the Tata Industries heavy industrial manufacturing group. Officials of the company professed ignorance of the problem but immediately offered help in organizing and conducting an intensive search and vaccination program throughout the greater municipal area. More than 2,200 cases were found. To keep the disease contained, all railway travelers were vaccinated before departure; bridges and major roads were barricaded. Only those who were vaccinated were allowed to pass. It took two months before the outbreak was controlled. Eventually, the government and Tata Industries reached an agreement for Tata personnel to assume responsibility for the program throughout southern Bihar state. From the Jamshedpur epidemic alone, 300 additional outbreaks and 2,000 cases had occurred in eleven states of India and in Nepal.

In the face of these problems, Srivastav decided that the only way smallpox could be brought under control was to cease all efforts directed toward surveillance and containment and to launch an all-out "backlog fighting program," with the goal of 100 percent vaccination coverage. This had been the failed strategy of the mid-1960s. Despite the protests of our program staff, Srivastav adamantly persisted. As problems mounted in Bihar and Uttar Pradesh, he insisted he would be thwarted no longer. Sharma learned of Srivastav's decision only when the Bihar minister of health began to take action in June to shift program staff. Sharma bypassed Srivastav and appealed to the minister of health, Karan Singh. Together they flew to Bihar to put the program back on track.

It was clear that the special surveillance and containment teams were critical, especially in Bihar and eastern Uttar Pradesh. Additional Indian and WHO epidemiologists were hurriedly recruited and assigned. The number grew from twenty-six in October 1973 to seventy-nine in June 1974. With added personnel, vehicles, per diem, and other costs, our resources were stretched to the breaking point. Appeals for support from donor governments were not successful. Skeptical officials politely refused to provide further assistance. Their misgivings were understandable: the number of smallpox cases in India during the spring was the highest recorded in more than twenty years. Moreover, the donor countries were regularly being solicited by frequent and urgent WHO appeals to bolster the foundering malaria program.

I spent much of my time through the spring of 1974 trying to raise more support, but with little success. Just as I was about to give up, I learned from Dr. Grasset that Swedish International Development Authority (SIDA) funds might be available (they had originally been ear-

---

THE INVALUABLE IMPREST ACCOUNTS

Indian Central Appraisal Team members rated the introduction of imprest accounts as one of the most important initiatives of the program. It was a simple but novel concept for countries of the region. Administrative officers gave team leaders money in advance, sufficient to meet their expected reimbursable expenditures during the forthcoming two to four weeks. At the next meeting, the leaders provided receipts for what they had spent, and the imprest account was restored.

Until this system was established, there were many vexing problems that took time to resolve, if they could be resolved at all; for example, the repair of a vehicle would require weeks of paperwork while the vehicle sat idle; hiring of temporary staff to serve as watch guards at infected houses was impossible; provision of office supplies was a continual problem.

Some of the expenditures were out of the ordinary, and special justifications were needed. There are several that I remember well. One was a receipt for costs associated with rental of elephants—to ford rivers and, later, for public advertising of a reward for reporting a case. Food costs to feed beggar families in infected compounds—so that family members would not go out begging. Purchase of the body of a deceased smallpox case for burial—otherwise it would have been floated down the Ganges. A series of quite large expenditures allegedly for car maintenance and repair that had, in fact, purchased a tank car load of gasoline at the time of the oil embargo in order to keep the vehicles on the road.

marked for a project that had been canceled). Grasset was eloquently convincing that the smallpox program was a worthy beneficiary. Within weeks, a memorandum of agreement was signed providing $2.8 million. Over time, Sweden would provide $15 million in support of the program. We received other invaluable help as well from the CDC. WHO's administrative staff had been overwhelmed by the burgeoning number of personnel in the field and the increasing expenditures for petrol, vehicle repair, per diem costs, printing, shipment, and other items. Knowledgeable management staff was needed. As he had done before, Dr. David Sencer, chief of the CDC, immediately came to the rescue, sending some of the CDC's principal administrative officers, including the center's deputy director, William Watson.

A SUMMER PROGRAM—1974

Early June 1974 was the psychological nadir of the Indian smallpox eradication effort and perhaps of the global program itself. For nine months the intensive campaign had been in progress throughout India, with a greatly increased field staff working frantically to control outbreaks. But by June there were still 8,700 active outbreaks that we knew about, and there were undoubtedly others. In all, 116,000 cases had been recorded— more than had been tallied worldwide during any of the program's previous six years! The staff was exhausted, and the weather was grueling with temperatures routinely over 40°C (104°F).

The epidemics in India made world headlines. On May 18, India tested its first atomic device, an event extensively covered by the international press. Meanwhile, local headlines throughout India reported the grim news about smallpox: more than 11,000 cases had been reported in a single week. The international press highlighted the paradox: sophisticated technological achievement in a country not yet capable of dealing with a disease that most countries had now conquered. Another important piece of news broke about this time—Pakistan expected to record its last smallpox cases in August (the actual date turned out to be October). Officials in India considered the prestige of the country to be at stake.

In a June meeting with the secretary of health and Director-General Srivastav, the Indian government agreed to fund an emergency program.

The number of epidemiologists was to be increased to one hundred. Also, 300 additional containment teams would be recruited—each to be headed by a recent Indian medical graduate. Meanwhile, the chairman of the Tata group, Mr. J. R. D. Tata, approved expenditures of $900,000 for the provision of personnel and vehicles. Tata's endorsement carried important political weight; he was a major supporter of the Congress Party and had personal access to Prime Minister Indira Ghandi.

Search programs were intensified throughout India, strongly encouraged by a public statement from the prime minister, requesting "the fullest cooperation of all citizens." Progress was to be monitored in terms of the number of infected villages—specifically, any village that had recorded one or more cases of smallpox within the preceding six weeks. At the end of June 1974 there were 6,400 infected villages. Elaborate measures were taken to ensure that every house with a case would have twenty-four-hour guards to make certain that the patients did not leave the house and that anyone who entered was vaccinated. All residents living within the nearest 500 to 1,000 houses would be registered and vaccinated as well. Poor families received a stipend to cover food costs during the quarantine period. Monthly searches were conducted in Bihar and Uttar Pradesh and a reward was offered to anyone reporting a case. By November 1974 the number of infected villages had fallen to 340. However, no one could forget that in January, only eleven months earlier, the situation had also appeared encouraging—until smallpox epidemics exploded across Bihar.

Disasters mounted. In August and September the worst floods in twenty years ravaged Bangladesh. Their aftermath was famine and tens of thousands of new refugees. Smallpox began spreading rapidly, the most heavily affected areas being along the northern Bangladesh-Indian border. Special surveillance teams repeatedly searched these areas, eventually discovering thirty imported outbreaks. Another setback was the discovery of forty infected households at one of India's holiest sites, Bohd Gaya, located near the capital of Bihar. Pilgrims had gathered to celebrate the 2,500th year of the death of the religious leader of the Jain sect. Some of this sect traditionally resisted vaccination. The principal religious leader reluctantly agreed to vaccination. Guards were stationed day and night to watch each infected house; the entire area was quarantined by military police. Five adjacent villages were infected, but the last case occurred in February.

In April 1975, an army of 115,000 health workers conducted a week-long house-to-house search throughout India. Independent assessment of 5 percent of the villages showed that 85 to 95 percent of all houses in each district had been searched. To spur reporting, a reward was offered to anyone who reported a confirmed case and also to the health worker who validated it. Posters were distributed throughout the country (see figure 51). Only a few imported smallpox outbreaks could be found. On May 18, a case was discovered on a railway platform in Assam State where trains stopped en route to eastern India. The patient was a thirty-year-old homeless beggar. During the four days she was there, 4,500 railway tickets to seventy different stations had been sold. This prompted an intensive widespread search, but there were no more cases. She was the last.

On June 30, only thirty-five days after the last patient had been isolated, Minister of Health Karan Singh announced that smallpox had been eradicated from India. There was to be a public celebration on August

Figure 51. Poster of a Hero Slaying the Smallpox Demon. Posters advertised that a reward would be given for anyone reporting a case confirmed as smallpox. In this poster, the hero is slaying the smallpox demon with a bifurcated needle. *Photograph courtesy of WHO.*

15—India's Independence Day. It was headline news in India and around the world. However, our entire staff was deeply concerned that the announcement might be premature, and this we did not want! I recalled only too well other occasions when many weeks had passed after we thought we had found the last case in a country—only to discover later that the excitement was premature, calling into question the credibility of WHO. The circumstances of the last case, the vast areas of India where small pockets of smallpox might still persist, the problems of fearful health officers hiding cases—all of these factors dictated the need for a discreet silence until far more time had passed. However, the minister was elated and could not resist announcing the news to the public.

We anxiously awaited Independence Day, the most important national holiday in India. National flags were flying everywhere and special flag-raising ceremonies, parades, and cultural events were taking place throughout the country. The heart of the celebration was in New Delhi where the high point was a special address by the prime minister. For me, it was an unforgettable moment when Prime Minister Indira Ghandi saluted the people of India on their twenty-eighth year of independence and proclaimed that India, for the first time in its long and storied history, had won freedom from smallpox. Fortunately, no further cases of smallpox were found.

Following the ceremonies, Director-General Mahler and I rode to the airport expecting to take a plane to Dacca, Bangladesh—the last endemic country in Asia and the last country in the world with the severe form of smallpox, *variola major*. The plane did not take off. All flights to Bangladesh had been canceled. The country was in chaos. The president of Bangladesh and his family had been assassinated and martial law had been declared. Indian troops were preparing to move to the border, bracing for a new tidal wave of refugees, some of whom almost certainly would be infected with smallpox.

SMALLPOX IN NEPAL

Nepal was an entirely independent nation with its own problems with smallpox, but until 1971 our primary focus had been on India. Nepal was all but inseparable from India; 90 percent of its 12 million people resided within sixty miles (one hundred kilometers) of India's two most problematic states—Uttar Pradesh and Bihar. Its people were Hindu, and movement was entirely free along its long open southern border. It was one of the least-developed countries, closed to the outside world until 1951. Twenty years later its infrastructure was still in an early stage of development.

Until 1971 WHO provided little assistance for smallpox eradication except for freeze-dried vaccine. That year a unique program was launched under the direction of a new national director, Dr. Purushollam Shrestha and WHO adviser Dr. M. Sathianathan, a Sri Lankan. They were supported by two West Africa veterans, Jay Friedman and David Bassett. The plan was to conduct mass vaccination during one month each year using some 3,000 locally trained temporary vaccinators who would vaccinate more than 6 million people. During the other eleven months, the full-time staff of six hundred would cover their assigned districts on foot, seeking and containing outbreaks of smallpox. It was a new and surprisingly effective tactic in the eradication program.

By June 1972 it appeared that transmission had been interrupted. Then came a tidal wave of importations, first from Uttar Pradesh and then from Bihar. Over the following three years, Nepal was to record 1,900 cases in 240 different outbreaks—three-quarters of them were imported. Most were readily contained. Only about 25 percent of the imported outbreaks spread to other areas. On April 6, 1975, Nepal recorded its last case.

Smallpox eradication was the first national program in Nepal, the first to extend to all parts of the country, and the first to establish a national disease-reporting system. The remarkable accomplishments of this program further highlighted the pathetic performance of the two adjacent Indian states of Bihar and Uttar Pradesh.

Chapter 7

AFGHANISTAN, PAKISTAN, AND BANGLADESH

The Last Stronghold of *Variola Major*

Afghanistan, Pakistan, and Bangladesh each presented unique challenges, more difficult than I had expected (see figure 52). These three countries had the potential to be more manageable than either India or Indonesia, each being smaller in area and with far fewer people. I had hoped for an outcome similar to that in Africa. There, with ample supplies of freeze-dried vaccine, improved supervision, and fairly simple surveillance-containment operations, we were able to stop smallpox transmission in most countries within two to three years. Only Afghanistan, the least developed of the three, met expectations.

Pakistan and Bangladesh, like India and Indonesia, were densely populated. Inexpensive public transport by trains, buses, and boats (in Bangladesh) was comparatively plentiful and well used. People thus spread smallpox more frequently and widely than in Africa. Numerous vaccinators were employed in both countries, but their effectiveness was limited by poor management and archaic health structures. One illustration was the vaccination program. In Africa, vaccinators were expected to vaccinate 500 people per day, but vaccinators in Pakistan and Bangladesh usually averaged no more than five to fifteen per day.

When the global program began, Bangladesh was part of Pakistan and was designated as East Pakistan Province. Civil war broke out in 1971, and

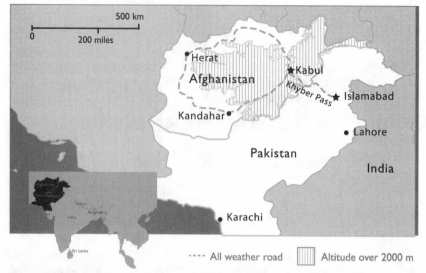

Figure 52. Afghanistan population (1970): 11,800,000.
Pakistan population (1970): 59,600,000.

the province became the independent country of Bangladesh. Success in eradicating smallpox in both countries repeatedly wavered in the balance and was uncertain even as late as 1975. Finally, in October 1975, the last outbreak of smallpox in Asia was discovered and contained on Bhola Island, Bangladesh.

Afghanistan was a wholly different story. Its government structure was rudimentary. It was in an early stage of economic and institutional development. Some areas remained tribal and did not recognize the central government. Vaccination was little known outside Kabul, the capital, and smallpox was believed to be widespread. The observance of *purdah*, which confined women to their homes, was common. These secluded women were said to be all but impossible to vaccinate, even with the permission of the head of household. Afghanistan seemed impregnable. Privately, I believed that Afghanistan might prove to be one of the world's most difficult problem areas, if not the last reservoir of smallpox. Much to everyone's surprise, Afghanistan's last indigenous case occurred in September 1972—well before success in Pakistan, India, or Bangladesh. This was due to a remarkable team of Afghan and international staff who overcame unbelievable obstacles.

AFGHANISTAN

Afghanistan is an isolated, landlocked country. In the early 1970s its population was estimated to be about 12 million, widely scattered over an area about the size of Texas. Five small cities accounted for about 10 percent of the population. The rest of the people lived in 20,000 small villages at altitudes of more than 5,000 feet (1,500 meters). Throughout most of the country, vaccination was unknown. We assumed that smallpox must be widely prevalent, but we had no way of knowing for certain. Only a few hundred cases were reported each year, but then only a handful of the health centers and clinics provided reports of any kind.

A single all-weather road circled the country. The few other roads were in deplorable condition. Security was uncertain and dependent upon local tribal chiefs. Communication systems were sparse. A potentially insoluble problem was the practice of the ancient technique of variolation as a means of preventing a more severe illness and possibly death from smallpox. Its practitioners took scab or pustular material and infected others by scratching it into the skin. Only 1 or 2 percent died by acquiring smallpox infection in this manner, compared to 30 percent among those infected by the usual face-to-face contact. The more serious problem, however, was that the variolated individuals could transmit smallpox to others who were susceptible. After a variolator visited a series of villages, it was not uncommon to find a chain of smallpox outbreaks along the route he had traveled. Any national attempt to outlaw the practice was meaningless because there was no government structure that could enforce it.

Afghanistan had begun a smallpox control program in 1963 with a WHO adviser and freeze-dried vaccine donated by the USSR. As many as 500,000 people may have been vaccinated in and around Kabul during the program's first five years—but no one knew for certain because the vaccinators were illiterate and unsupervised and no records were kept.

I visited Afghanistan for the first time in November 1968, accompanied by Dr. Jacobus Keja, the Southeast Asia Regional Office (SEARO) smallpox adviser. We found a sorry disaster. The senior WHO smallpox program adviser seldom left his office, preferring to communicate with his Afghan counterpart by letter as he advised him on how the program should be run. He shared his office with two recently arrived WHO nurses: a very tall Russian, Ludmilla Chicherukina, and a short, petite Burmese, Khin

Mu Aye. His contact with them was usually by written memos. Periodically, when the report of an outbreak was received, vaccinators were dispatched to the area on a flatbed truck. After a time—days to weeks—they filtered back to Kabul. There were no reports to indicate what the vaccinators had achieved or where.

The situation was appalling. We arranged for the senior adviser to be transferred out of the country and developed a new plan for the program. Recruitment began for a new adviser. Dr. Arcot Rangaraj (see figure 53), recently retired from the Indian Army, was selected. He was a highly decorated medical service colonel with remarkable skills in organization and management. The government appointed a full-time Afghan program director, Dr. Abdul Mohammed Darmanger, who was no less energetic and determined. Both would spend the next five years with the program. Vehicles, vaccines, and other equipment were ordered, and by autumn 1969, an effective operation was ready to begin.

The status of the program in 1969 was vividly described in the report of a joint Afghan-WHO inspection team—and this was six years after the fully supported WHO program had begun:

- The director has little apparent authority, no defined budget, no ability to establish or enforce personnel policies, no authority to deploy vehicles or personnel without written authorization by his superior.
- The national program office has no telephone, secretarial assistance, stationery, files, forms, or records.
- One vaccination team has worked for only one week in the preceding eight, and 13 of the 53 team members have now been sent into military service.
- A second team has spent the summer giving cholera vaccinations.
- No maps of any sort are available.

By summer 1969, as field operations began, Rangaraj and Darmanger had a staff that included three Afghan medical officers, three WHO advisers, a small group of US women Peace Corps volunteers, and 298 Afghani field staff. They set in motion a meticulously organized, closely monitored, and well-documented operation. Five-person vaccination teams were organized with planned schedules and a target of vaccinating

Figure 53. Dr. Arcot Rangaraj, Indian physician, he was WHO senior adviser in Afghanistan (1969–1974). Later he served as director of field operations in Bangladesh (1974–1975).

an average of one hundred people per day, per team member. After twenty-four days in the field, they were given a seven-day rest period. The target for vaccinations was lower than in Africa but had to take into account the travel time between widely separated villages and the inevitable delays in securing permission from the head of each family before vaccination was performed.

When the vaccination teams finished in an area, two-man assessment teams checked a randomly selected, 10 percent sample of villages. If fewer than 80 percent had been vaccinated, or less than 95 percent of the primary vaccinations were successful, the teams had to revaccinate the entire area. Within months the teams were achieving their goal of 300,000 per month. During the first three years, 10.5 million people were vaccinated. A second round of vaccinations, aimed at children less than fifteen years of age, was begun in 1972 and completed eighteen months later. At the end of this second round, scar surveys in each of twenty provinces found that between 91 and 98 percent of those under fifteen years had vaccination scars. In a country where health services of any type had previously extended to only a small proportion of the population, this was an extraordinary achievement.

It had been widely believed that the all-male smallpox field staff could not vaccinate women in purdah. Accordingly, a request had been made to the US government several years before to send a contingent of female Peace Corps volunteers who would be given responsibility for vaccinating the women. A courageous, adventuresome group had come and had worked hard. But they were few in number and could accomplish little in the earlier, poorly administered program. Under the new plan, supervisors preceded the teams. They sought out tribal leaders and heads of family to

LESSONS IN PERSONNEL MANAGEMENT

To work effectively in different cultures, it is essential to understand and adapt. But how to reconcile divergent practices?

At a smallpox meeting in New Delhi, an obviously troubled Darmanger came to me seeking advice. He said, "I am an educated man, a physician, the director of the program. I had two serious discipline problems and I'm concerned that I may not have handled them like a professional should have." I asked for the details. The first, he said, was with his driver. "I often came back from the field late in the evening and he was angry about having to work long hours. I told him it was his duty. One morning he didn't show up for work. I was told he had gone to see the minister of health." (At that time in Afghanistan, the principle of a chain of command was often ignored. If one had a problem, an appeal directly to the top was not uncommon.) Darmanger continued, "I took with me two of my biggest vaccinators and went to the ministry. I found the driver in the inner office, about to meet with the minister. My two vaccinators each grabbed an arm and held him. I spoke quietly to him— if you go into that office, I will kill you. He knew I meant it. He has been no trouble since."

The other problem had been his response to a decision by the Kandahar-based vaccination teams to go on strike. I asked Darmanger why they had gone on strike. He said, "They hadn't been paid for three months but they shouldn't have gone on strike. It is their duty to work for their country." I asked what he had done. "Well, I took the leader into the office and beat him." And, what happened, I asked. "They went back to work and they haven't made any further trouble."

explain about vaccination and to request their cooperation. They were surprised to discover that, with this approach, there was a generally favorable acceptance for women to be vaccinated by the male vaccinators. Cultural modesty, however, forbade vaccination on the upper arm, as was customary. Instead, the women were often vaccinated on the wrist or back of the hand as they stood in the doorways of their houses. The Peace Corps group shifted over to work with the assessment teams and to provide much-needed administrative help in the field offices.

At the time the mass-vaccination campaign began, surveillance-containment operations were started. This was one of the few programs in which mass vaccination and surveillance-containment began simultaneously despite what our WHO manual advised. A zonal surveillance team was established in each of the country's four zones. All health units were to report weekly, and all malaria workers were ordered to report cases. Health units, however, were found in only 69 of 326 subdistricts, and the malaria program covered only part of the country. Accordingly, the teams moved from village to village, talking with leaders and visiting schools in efforts to find cases. The teams became so well-known that villagers would sometimes send false reports of an outbreak, knowing that this would quickly bring a team to the area. More than 1,000 cases were documented in 1970, the first full year of surveillance (see figure 54).

Rangaraj and Darmanger were surprised to discover that fully one-third of all smallpox cases were in outbreaks that had been started by variolation. In some areas, more than half of all adults bore variolation scars. The scars differed from vaccination scars in that they were on the hand or lower arm and were generally much larger. Vaccination had not reached many of the rural areas; the only recourse for inhabitants there was variolation for preventive purposes. Most variolators were farmers or religious leaders whose fathers and grandfathers had practiced before them. The surveillance teams made a concerted effort to find as many of the variolators as possible, to learn of their methods, and to persuade them to stop, if possible. Finding them was often difficult because they feared government punishment if caught. But stopping them altogether was unacceptable until vaccination became more widely available. Many variolators, however, agreed to perform vaccination instead of variolation and were given vaccine to do so.

The variolators had used scabs or pustular material taken from a

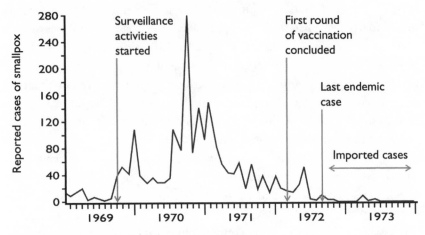

Figure 54. Afghanistan Reported Cases (1969–1973).

smallpox patient and kept in a container. They placed several scabs in a dish and added either spiced water or honey. This material was then scratched into the skin. Variolation was usually performed only during the snow-free months when foot travel was possible—thus, the question: for how long and where did the variolators store the scabs? Smallpox virus in scabs is very stable, especially if kept in a cool place. Interviews with ninety-seven variolators revealed that they did keep the scabs for many months, usually in protected, cool sites. However, all reported that for the

---

A SMALLPOX SURVEILLANCE TEAM—COURAGE AND DEDICATION

Under the direction of Darmanger and Rangaraj, the Afghani field teams became a remarkably dedicated and courageous group. The investigation of a rumored outbreak in a northern mountainous area is illustrative. The team was sent on horseback to investigate but encountered three-foot-deep snow and had to turn back. They tried again by another route but again encountered snow. The horses were abandoned and the team continued on foot for four days to get to the outbreak area. They moved from village to village vaccinating and checking for cases as they went. In all, they spent six weeks in the middle of winter containing the outbreaks. When it was possible to carry out a thorough search of the area in the spring, no subsequent cases of smallpox were found.

inoculations to be successful, they had to add fresh material each year. Some specimens were obtained from variolators that yielded smallpox virus when cultured in the laboratory, but it proved difficult to know for certain how long the material had been kept. Confidence that the practice had finally stopped would require at least two years of intensive surveillance following occurrence of the last case.

The number of cases decreased to 750 in 1970 and to less than 250 in 1971. The last endemic case occurred in September 1972, just thirty-six months after the program had begun—a remarkable achievement in a country so difficult.

There was no time for rest or celebration. Smallpox was still endemic in Pakistan, and tens of thousands of nomads annually migrated north into Afghanistan. The number of surveillance teams in southern Afghanistan was increased so that nomadic groups could be intercepted en route and vaccinated. Three separate importations occurred, but they were quickly discovered and contained. No evidence of successful variolation after 1973 was discerned.

PAKISTAN (WEST PAKISTAN PROVINCE BEFORE DECEMBER 1971)

When the global program began, Pakistan was a country consisting of two provinces, West Pakistan and East Pakistan, separated by more than 1,000 miles (1,600 kilometers) across India. Each province had an independent health service. In December 1971, after a civil war, East Pakistan Province became Bangladesh, and West Pakistan Province became known simply as Pakistan. This section deals only with smallpox in the area that was formerly known as West Pakistan.

The western part of Pakistan shared a long border with Afghanistan, but the two countries had little else in common. West Pakistan's 60 million people lived principally in the vast Indus River plain and in the low foothills of the Himalayas. Only 10 to 15 percent lived in the remote mountains and desert areas. The country had a comparatively extensive infrastructure of trained health personnel, communications, and road, air, and train service. Early in 1968 Pakistan signed an agreement with WHO to undertake an eradication program, and a WHO adviser was assigned by

WHO's Eastern Mediterranean Regional Office. So began one of the more confusing, costly, and ineffective of all national programs.

The plan developed for West Pakistan by the adviser and the government called for a staff of 1,400 including twenty-two medical officers and a three-year mass-vaccination program. A facility would be built to produce freeze-dried vaccine. Teams of assessors would check on the work of the vaccination teams. Nothing was said about a surveillance program. Through 1971 the program's budget steadily dwindled, and poorly organized pilot vaccination programs accomplished little. The number of reported cases tripled, from 1,800 to 5,800. This sharp rise was not a reflection of better reporting. Nothing had been done to improve reporting.

Meanwhile, particularly informative studies of smallpox epidemiology had been started in 1966 by Pakistani epidemiologists at the Pakistan Medical Research Center in Lahore, working with a group from the University of Maryland, led by Drs. Tom Mack, Gordon Heiner, and David Thomas. Over the course of a year, they studied the pattern of urban smallpox transmission in the city of Lahore and in a large, nearby rural district. Their findings were highly pertinent to the national program. They found that population immunity was surprisingly high despite poor supervision and inferior vaccine, and that more than 85 percent of the population had vaccination scars or pockmarks. They learned that only a small percentage of the villages was infected at any one time even at the seasonal high point for smallpox and also that smallpox outbreaks usually died out spontaneously in rural areas. During the summer, when transmission was at its lowest, it persisted only in urban centers.

The investigators concluded that the best strategy for Pakistan was to abandon plans for mass vaccination because the vaccination immunity was already high. The surveillance-containment program should be the highest priority. It was also apparent that special efforts should be made to stop transmission in urban areas in the summer when cases and chains of infection were the fewest. This, in turn, should have a major impact in decreasing the number of smallpox cases in the autumn and winter.

Unfortunately, papers from these studies did not become available until late in 1970, and most were not published until 1972. Although they were then widely distributed and presented at meetings, little notice was taken of them either by the Pakistan leadership or by WHO advisers.

The country's administrative structure deteriorated sharply in 1971 coincident with East Pakistan becoming the independent nation of Bangladesh. The government divided West Pakistan into four provinces and two federally administered regions. Each province was given total autonomy in health matters. The Pakistan national administrative office remained intact but made no effort to communicate with provincial programs, to coordinate their efforts, or to oversee their activities. It was an untenable situation: national authorities insisted that all WHO contacts and program planning must go through them—and that they alone would represent Pakistan at international conferences. The now-autonomous provinces were further divided into ten to twenty districts, each with a generous complement of supervisors and vaccinators. Further confusing the situation were municipal corporations that functioned independently with their own personnel and were not accountable to other medical authorities. The only national coordinating links were WHO advisers. This administrative chaos blocked development of the intended vaccine production facility. No agreement could be reached as to whether the central government or the provinces should pay for the building and the operating costs. Vaccine production equipment, received from UNICEF, rusted untouched in a warehouse.

In 1973 three events heightened the level of governmental interest. First, the reported number of smallpox cases rose to 9,300, the highest total since 1948. Next, Afghanistan announced that it had become smallpox-free except for importations from Pakistan. This was an embarrassment to Pakistan's national pride. The third event was India's decision in the summer of 1973 to mobilize its vast resources of health workers to undertake monthly village-by-village searches with the expectation of stopping smallpox transmission within a year.

Pakistan agreed to our proposal that a WHO adviser be appointed for each of the four provinces. A marked transformation in the program began with the arrival of Dr. Omar Sulieman, the dynamic former director of the Sudan program; Dr. P. R. Arbani, who had directed surveillance in Indonesia; and Dr. Reinhard Lindner, an Austrian WHO adviser who had worked in Indonesia. Finally, a knowledgeable general from the Pakistan Army Medical Corps was assigned to provide some semblance of national direction and coordination.

Given the high levels of vaccination immunity, it was decided in January 1974 to suspend the mass-vaccination program, to redirect the vaccinators and supervisors to surveillance-containment operations, and to conduct village-by-village searches. This was similar to the new strategy in India. A reward was announced for reporting a case, and rigorous containment measures were implemented.

On September 28, 1974, the last-known case of smallpox was detected in a village in Sind Province. There were no celebrations, however. The surveillance system was still so inadequate that months would elapse before the staff could be confident that the last chain of smallpox transmission in Pakistan had finally been broken.

BANGLADESH: THE END OF *VARIOLA MAJOR*

Bangladesh was the last country in Asia with smallpox (see figure 55). It was also one of the poorest, most densely populated countries in the world. Ironically, Bangladesh (East Pakistan until December 1971) had managed to stop smallpox transmission five years earlier—long before India, Pakistan, or Indonesia had. But in 1971, civil war erupted and 10 million refugees fled to India—250,000 of them were housed in the Salt Lake refugee camp near Calcutta. Smallpox broke out in the camp. In December 1971 Bangladesh declared its independence and the refugees streamed home, taking smallpox with them. If the authorities in charge of the camp had ensured that everyone had been vaccinated—as officials at other camps had done—they could have spared at least 200,000 people from smallpox, 40,000 of whom died.

From January 1972 until the last case in October 1975, a competent but increasingly frustrated and discouraged staff dealt with incessant crises. As each seasonal epidemic drew to a close, and the June monsoon rains began, the staff felt a surge of hope—believing that this time they could surely stop transmission within the following few months, during the usual seasonal low point in incidence. But one unexpected tragedy after another rudely dashed these hopes. In August 1975 came yet another disaster, which threatened the entire program. In spite of this, the transmission of smallpox was stopped—finally—before the 1975–1976 epidemic smallpox season began. If the effort had not succeeded then, I wonder today whether a battered staff and WHO's exhausted fiscal resources could have sustained the momentum in Bangladesh and, indeed, that of the global program.

A richly fertile country, Bangladesh lies in the delta of three great

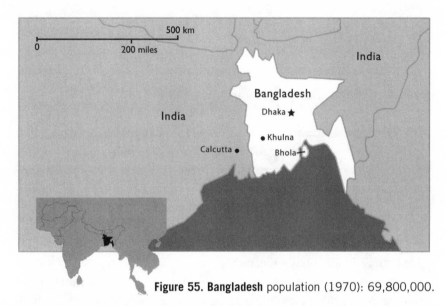

Figure 55. Bangladesh population (1970): 69,800,000.

Asian rivers. Most of the land is only fifty feet (fifteen meters) above sea level; at least one-third of it is flooded each year during the monsoon season. In the early 1970s, 95 percent of Bangladesh's 70 million people lived in rural areas—but nearly a third of the population worked part-time on other farms or in the cities. Large numbers of laborers migrated around the country to find temporary work, especially during the floods. In this environment, smallpox spread rampantly.

A serious eradication effort had begun in 1962. Over a three-year period, 72 million people were vaccinated; 68 million more were vaccinated over the next three years. Both numbers were larger than the total population at the time. However, vaccinators had used a wholly unsatisfactory liquid vaccine produced locally. Given the high temperatures in the country and the poor stability of the vaccine, it is likely that many of the vaccinations were unsuccessful. Several thousand cases of smallpox continued to be reported each year—just as they had before the campaign started.

In 1967 the Pakistan Medical Research Center undertook a one-year study of smallpox transmission in a rural area. The study was similar to that conducted in West Pakistan, and so were the results. A surprisingly high proportion—81 percent—had vaccination scars. Most of the outbreaks were small, did not spread readily, and were self-limiting. The source of the outbreaks tended to be the large cities. As in West Pakistan, the study provided the evidence needed for a strategy that focused on surveillance-containment in urban areas. However, the program was already set on course for another mass-vaccination campaign.

A renewed eradication program begins—1968

The renewed eradication effort called for a three-year mass-vaccination campaign, but it also provided for four teams that would be responsible specifically for controlling smallpox outbreaks. Progress was slow. Government funds were not approved until April 1969, and over the next fourteen months, only 4.5 million people were vaccinated by the teams. At the same time, an army of local vaccinators reported vaccination totals of 30 million people per year—but supervision was negligible, and these figures were believed to be highly inflated.

I was especially concerned about the vaccine laboratory, which began production in 1967. Plans called for it to make enough vaccine for both East and West Pakistan, but it never succeeded in reaching half of this target. Much of the vaccine was of low potency. The laboratory director claimed that its vaccine met WHO required standards, but confirmatory tests performed by the WHO Reference Center in the Netherlands found that fewer than half of the batches were satisfactory. The director responded angrily, insisting he would send no more specimens for testing. Unexpectedly, four samples arrived a few months later, and all were of good quality. This was encouraging but seemed too good to be true. When I visited the laboratory, I was taken aside by Dr. Ataur Rahman, the deputy director of the laboratory, who told me what had actually happened. All vials, he explained, were from a single batch that had previously been proved satisfactory by the reference laboratory. They had been relabeled to make it appear that they had come from four different new batches. The laboratory director unhappily accepted the assignment of a WHO consultant to work with him, and Rahman assumed greater responsibility for production and for selecting vials for independent testing. Production increased, and by 1973 the laboratory was producing 20 million acceptable doses each year.

In 1968 the number of smallpox cases rose to more than 9,000. A concerted surveillance-containment effort was required, but little was done. In November 1969 we organized a seminar in Dhaka and brought in staff from Latin America and West Africa to report on their successes with surveillance-containment. Surprisingly, the seminar had the desired effect. The staff decided that from January 1970, every case would be investigated and containment measures taken. Physicians were designated to direct programs in the country's four divisions. Each division was to have five teams of ten supervisors and vaccinators. My deputy, Dr. Isao Arita, spent the months of February and March working to set up the program. The mass-vaccination campaign was suspended, reporting was streamlined, and malaria workers, who visited all houses in Bangladesh every two weeks, were asked to report all smallpox cases they found. In April 1970 only 260 cases of smallpox were detected, and these were in only four of the nineteen districts. By August the last outbreak was contained. Intensive searches continued until March, throughout those months when smallpox was usually most prevalent. No cases could be found.

Civil war—March 1971

Civil war broke out in March 1971 and lasted until December. Between 1 and 3 million civilians died, 10 million fled to India, bridges were blown up, more than a million houses were destroyed, and the health structure was severely crippled.

Until December 1971 when the refugees began to return, no rumors of smallpox surfaced and no cases of smallpox were registered at any of the hospitals. It was difficult to believe that the country was free of the disease— even though no cases had been found among the refugees entering India. In July 1971, in the midst of the civil war, I was asked to lead a small team to Dhaka on behalf of the United Nations—primarily to negotiate bringing in foreign humanitarian aid officials whose presence might serve to diminish the extensive violence. I met privately with the four provincial surveillance team leaders, M.A. Sabour, M. Yusuf, M. Shahabuddin, and Dr. M. Aftabuddin Khan. Somehow, these dedicated physicians had continued their surveillance work in Bangladesh throughout the civil strife, managing to travel extensively, constantly searching for smallpox cases. They were not only unpaid—they also were obviously taking great personal risks. As evidence of the breadth of their travels in search of cases, they provided remarkably detailed information as to which roads were accessible, which bridges were out, which areas were mined, which ferries were running, and where the fighting had been most severe. Of significance was the fact that they had been unable to find a single case of smallpox.

By December, sixteen months had elapsed during which no evidence of smallpox had been found in all of Bangladesh. The ease and rapidity with which transmission had been interrupted was a stunning surprise and generated a degree of confidence and optimism that turned out to be unwarranted.

Reinfection—December 1971

Refugees that had fled to India and entered the temporary refugee camps were screened for possible smallpox and vaccinated. None had smallpox, but, as noted earlier, smallpox from neighboring Indian villages had been brought into one camp, the Salt Lake camp near Calcutta. Misdiagnosed as chicken pox, it soon became an epidemic and was not accurately diagnosed until December 19, three days after Bangladesh had been declared an independent country. Refugees streamed back across the border. Obvious cases, people in the incubation period, and unvaccinated contacts were loaded together on trucks and trains for the trip. It was the season when the transmission of smallpox was rapidly increasing. And with 27 million displaced people moving from place to place, herded together in temporary camps and city slums, smallpox spread explosively (see figure 56).

Trying to stop an epidemic at the peak of seasonal transmission was hard enough. Other problems surfaced—half the motor vehicles in Bangladesh were damaged or had disappeared, as had two-thirds of the bicycles. Boats had been damaged or sunk. The civil administration, including the health services, was a confused shambles. Arita immediately flew from Geneva to Dhaka, and four WHO epidemiologists were hastily recruited, including Dr. Stan Foster, former senior smallpox adviser in Nigeria, and Dr. Nilton Arnt, an epidemiologist from the Brazilian program. They were soon joined by others who were to remain with the program on long-term assignment: Dr. Nick Ward, from the health service in Botswana, and Dr. Daniel Tarantola, from service with a French volunteer medical group in northern Bangladesh.

Figure 56. Refugee Camp, Bangladesh. From 1971 to 1975, millions of refugees were accommodated in densely crowded camps as civil war, major floods, and famines devastated the country. *Photograph courtesy of UNHCR.*

The epidemic spreads—1972

It took heroic efforts to reinvigorate a disheartened health staff and to control smallpox in the refugee camps by searching for cases and vaccinating. The team hired 3,800 temporary vaccinators, and the United Nations Disaster Relief Operation provided motor vehicles, boats, and bicycles. By March 1972 the program had begun receiving weekly telegraphed reports from many of the fifty-seven subdivisions of the country.

At first, major outbreaks were confined to areas near the Indian border, but, throughout 1972, smallpox continued its march. By the end of June, 6,000 cases had been reported (see figure 57). Almost all—95 percent— were from only four of the nineteen districts. The Pakistan Medical Research Center's studies had shown that the great majority of new outbreaks were imported from urban areas and that rural outbreaks tended to stay relatively confined and to spread slowly. This was not the case in Bangladesh this time.

Special attention was paid to Dhaka, the largest city, where four-

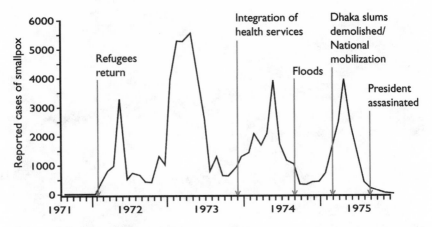

Figure 57. Bangladesh Reported Cases (1971–1975).

person surveillance-containment teams were assigned to work in the most heavily infected areas. Outbreaks in Dhaka began to develop in October 1972 in slum areas and resettlement camps. In December, 200 cases were detected and the program staff braced for a major effort. Eighteen mobile surveillance teams were formed, and checkpoints were established at railway and bus terminals to vaccinate passengers. Mass-vaccination campaigns were conducted day and night in slum areas and camps. During the 1972–1973 winter season, 1.7 million people were vaccinated in Dhaka alone. Despite these efforts, case numbers steadily increased, reaching nearly 6,000 in May, and smallpox spread to many rural areas.

With the monsoon rains, the epidemic abated significantly, but still some 400 to 600 cases per month were being detected. What was disturbing to me and to the staff was that an enormous effort had been made, with a primary emphasis on surveillance and containment, yet smallpox had not been responsive as it had been in Africa, nor in pre–civil war Bangladesh.

In October 1973 the situation brightened. Only 400 cases were reported, the lowest monthly total since the disease had been reintroduced. Dhaka was once more free of smallpox, and the staff estimated that infection remained in only 150 villages. A new system for charting progress was adopted—the units being the "infected villages." A village was deemed infected until six weeks had elapsed since its last case.

The disaster of November 1973 and recovery

Another setback: to the consternation of everyone, the government suddenly and inexplicably decided in November to suspend all health activities for three months to reorganize the health services. The large and previously autonomous malaria program field staff, as well as all other village
health workers, would be retrained to be "family welfare workers." Their
tasks would be to register married couples and births, perform smallpox
vaccinations, search for cases of malaria and smallpox, distribute vitamin
A capsules and contraceptives, and provide health education and family-
planning materials. After the training period, they were instructed to spend
two more months on paperwork—preparing a separate health card for
each family, listing the names and ages of each family member.

This devastating change in operations occurred at the start of the
smallpox season and it took the staff months to reestablish the eradication
program. All of the progress that had been apparent in October evaporated. The number of infected villages jumped from about 100 to nearly
1,000 in April 1974 as outbreaks exploded across Bangladesh, most of
them in the northern part of the country.

Drastic measures were called for. The government issued an "Emergency Plan for Smallpox Eradication under the Integrated Health and
Family Planning Program" that unified all health services under one central authority. Dr. Mahboober Rahman (see figure 58), a tough, capable
former director of malaria eradication, was given overall responsibility.
Additional epidemiologists were brought in, as well as field radios for
better communication. There was hope again with twelve WHO epidemiologists and 10,500 family welfare workers in a reorganized and better-
defined structure, and with smallpox eradication a priority, that during the
autumn 1974 monsoon season—the low point in smallpox transmission—
the earlier success of 1971 surely would be repeated.

By the end of October 1974 the number of infected villages had fallen
to just ninety. Victory seemed virtually within reach. We were excited, yet
apprehensive that the goal might again slip away. Beginning about this time
and for the following two years, I became a perpetual itinerant, commuting
from Geneva to the remaining critical areas of northern India, Bangladesh,
and Ethiopia, traveling for two to three weeks, returning for a week to deal
with other global program issues, and then returning to the field. I could

Figure 58. Dr. Mahboober Rahman, national director of the integrated health program from April 1974. A former director of the national malaria eradication program, Rahman brought order to a chaotic administrative structure. Dr. Daniel Tarantola, WHO adviser, is seated at the side. *Photograph courtesy of WHO/L. Dale.*

not decide whether a dry 44°C (110°F) summer in Bihar (India) was more or less uncomfortable than 35°C (95°F) and 100 percent humidity in Dhaka. Everyone was working seven-day weeks, ignoring personal discomfort. By August, smallpox in Pakistan was virtually gone (its last cases were in October). This meant that smallpox in all of Asia was essentially confined to only one state of India and two districts in northern Bangladesh.

The staff, while optimistic, was near exhaustion as they desperately sought to contain the last outbreaks. But further disasters lay ahead.

Bulldozers, floods, and famine trigger yet another disaster—1974

The 1974 monsoon rains were heavy; the most severe floods in two decades struck the northern areas of Bangladesh, the center of the outbreak areas. The extensive destruction of crops brought famine and tidal waves of refugees, many settling in camps and slum areas near Dhaka. The number of infected villages rose from 90 in October to more than 200 in December, and then came a second tragedy. The government decided to bulldoze many acres of slums around Dhaka—places where refugees had temporarily settled and where smallpox was now burgeoning. It was a disastrous decision: 100,000 people, now homeless, dispersed to other areas across the country. Smallpox began erupting everywhere. Fieldwork was difficult and could be treacherous (see figure 59). It was a nightmare recurring.

Additional epidemiologists, vehicles, and motorcycles were required, as well as expenses for travel, gasoline, and other equipment. We needed funds urgently. However, we had already exhausted our own resources, as well as the discretionary funds of WHO's regular budget in dealing with the crises in Bihar and Uttar Pradesh in India. We would have to seek additional international support. However, this necessitated the formal approval of the Bangladesh Planning Commission, and their approval was contingent upon the submission of a revised and detailed plan of operations. After that, some time would be required for a decision. We couldn't afford the delay. Special measures were needed to bypass the system. Fortunately, a new principal health adviser to the planning commission had just been appointed—Dr. Ataur Rahman, the physician who had rescued the vaccine production laboratory. A highly regarded scientist, he persuaded the commission to issue an immediate direct appeal. The Swedish International Development Agency responded promptly, as it had in India, and made available $3.5 million. Contributions also came from Canada, Denmark, Norway, and the United Kingdom.

By February 1975 the number of infected villages had nearly quadrupled; more than 2,000 cases were reported in that month alone. This led to a presidential directive declaring smallpox to be a national emergency and ordering the mobilization of all available resources. The number of epidemiologists grew from fourteen to forty, many of whom were veterans of service from elsewhere in Asia and Africa. Dr. Rangaraj, the senior adviser in Afghanistan, was brought in to direct field operations; thirty-five Indian-

Figure 59. A Footbridge in Bangladesh. Bangladeshis were accustomed to single-pole bamboo bridges, but foreigners were not. Alan Schnur joined the program as a Peace Corps volunteer in Ethiopia and later worked in Bangladesh. *Photograph courtesy of WHO/P. Roberts.*

made Jeeps were purchased and driven to Dhaka; volunteers from the British charity OXFAM established a workshop/motor pool for a fleet of more than one hundred Jeeps; a large jute-and-fiberglass emergency operations hut was built by the US charity CARE. Administration and financial management were a real challenge for a program that had been accustomed to spending $12,000 per month—but by February was spending $125,000 a week. Andrew Agle, a US veteran of western Africa and Afghanistan, and a Bangladesh administrator, Alim Mia, kept the resources flowing and saved the program from collapsing under a mountain of paper.

Meanwhile, national searches had been conducted in October and December 1974. Beginning in April 1975, these intensified and were conducted every four to six weeks. Between May and July, 112,000 people, on average, were working in the field every day. A 100-*taka* reward was offered for the report of a case.

A greatly expanded radio network eased the communications problems. Outbreaks were more thoroughly investigated and the sources of infection traced. With the help of radio communication, more than one hundred sources of infection each month were reported to other areas and to India. Ever-more-stringent methods were used to isolate patients: round-the-clock house guards, the complete vaccination of everyone within a half mile of the patient's hut, and the provision of food and water to the patient's family to make sure they remained in place and isolated.

Throughout this period, there was continuing tension between the Bangladesh secretary of health, a former surgeon, and the senior program staff, including both the WHO advisers and Mahboober Rahman. The secretary was not convinced that surveillance and containment could be effective, and he argued incessantly for mass vaccination. Certainly, the experience of the preceding three years did not make a convincing case for surveillance-containment, but mass vaccination in south Asia had proven repeatedly to be ineffective. The fundamental impediments, as we saw them, were the extraordinary density of the population and the extensive movement of migrant laborers and refugees and those displaced by the bulldozing of the slums of Dhaka. The only feasible answer was to greatly

---

SACKS OF MONEY: PROVIDING FUNDS TO FIELD STAFF

Providing transport and accommodation for the abruptly expanded temporary supervisory staff was a challenge. They required funds for petrol, emergency vehicle repair, lodging, and food for themselves and sometimes for families of patients being kept in isolation. There were no credit cards at that time, and branches of banks in peripheral areas were not yet functioning. The only practical solution was to provide all field supervisors with imprest accounts—basically an advance payment of funds—and for this to be replenished when they presented receipts at monthly planning and strategy meetings. However, the largest bank note then available was a 10-*taka* note (worth about 70 cents). Amounts of as much as 10,000 *takas* or more were sometimes needed for disbursement. Handling the large bundle of banknotes was often solved simply by putting the notes in a large burlap sack and tying this to the back of a motorcycle. Although no attempts were made to conceal what was being carried, no thefts ever occurred.

strengthen containment and to improve the search programs in order to detect outbreaks at the earliest possible time.

The secretary's frequent visits to the field were incredibly disruptive. Unannounced, he would travel to a district or town. After an impromptu speech lasting three or four hours, he would demand that health staff stop all activities and vaccinate everyone in the area. He threatened that if he found a single unvaccinated person after seven days, severe punishments would ensue. He also asked senior staff to pledge to resign if a single case of smallpox was found in their jurisdiction after a certain date. This blustering served only to suppress reporting. The secretary was accompanied on all his visits by the WHO country representative, a physician who knew nothing about smallpox or the program and surprisingly little about public health. He vociferously supported the idea of mass vaccination and whatever else the secretary suggested. This WHO representative had an additional liability—a notorious social life at odds with expected diplomatic behavior. I decided that an early replacement should be found. In April I flew back to Geneva and met with Director-General Mahler. He promptly summoned the representative for immediate discussions and transferred him out of the country.

By April 1975 the number of infected villages had grown to 1,300. There were seventy international epidemiologists in the field now, joined by national epidemiologists from Bangladesh medical schools and some voluntary organizations. The intensity of the program and of the effort being made was far greater than had ever been conducted in any country. At this time it was apparent that India was containing its last few outbreaks. We all realized that, somehow or other, *Variola major*—now cornered in this one small country—simply had to be conquered as soon as possible, lest it escape and reinfect one or more neighboring countries. The number of infected villages dropped precipitously—from 950 in May to 130 in July. The program intensified with the announcement that on August 15, India would celebrate its Independence Day and its freedom from smallpox for the first time in its recorded history. Bangladesh was now the lone endemic country in Asia.

ONE LAST DISASTER—1975

Director-General Mahler and I planned to arrive in Dhaka on the morning of August 16, after the ceremonies in New Delhi. We planned to meet with the staff to offer encouragement and to see what else could be done to help them reach zero cases at the earliest possible date. On arrival at the airport in New Delhi we were informed that all flights to Dhaka had been canceled. Sheikh Mujibur Rahman, the first president of Bangladesh, and his family had been assassinated. Martial law had been declared, radio communications were suspended, and India had begun preparations to deal with still another unanticipated inundation of refugees.

Surprisingly, Bangladesh remained comparatively calm, but for nearly a month, national and WHO staff had to limit their travel. However, the number of outbreaks continued to drop, and on September 14 the last-known smallpox case was discovered and isolated. Six weeks passed, and by late October there were no known infected villages anywhere in Bangladesh. All agreed, however, that caution was in order. We decided to wait at least another two weeks before making an announcement that smallpox had been banished from Asia.

House-by-house searches continued until early November when civil disorder erupted in three parts of the country. The United Nations recalled its personnel to Dhaka; only the WHO epidemiologists remained in the field.

On November 14, WHO announced at a specially convened press conference that two months had passed since the onset of the last case and that *Variola major* was now consigned to history. Jubilation lasted less than twenty-four hours. The next day a cable arrived from the densely populated Bhola Island, off the southern coast: "One smallpox case detected Village Kuralia. Date of attack October 30. Containment started." What with the repeated searches and the widely publicized reward, this had to be a misdiagnosed case—or so most thought. Foster himself went alone to check. It was not an easy trip, requiring nearly a day. He arrived to find the three-year-old patient hiding under a jute bag. It was smallpox. A call brought Tarantola and surveillance teams by speedboat, steamboat, Jeep, motorcycle, and finally by foot. It was November, the monsoon was over, and the season of highest transmission was beginning.

Bhola Island had a population of 960,000, living on approximately

1,000 square miles (2,600 square kilometers) at the mouth of the Ganges. Smallpox outbreaks had been discovered there during the monsoon period by the medical officer, but he had not reported them. The surveillance teams eventually found that 144 cases had occurred since the beginning of August, the last being the patient Rahima Banu, who had become ill on October 16 (the cabled information was incorrect). Additional staff came to Bhola to apply intensive containment measures. The patient was isolated and house guards were posted; anyone who had not recently been vaccinated was not allowed to visit. In all, 18,000 people living within about a one-mile (two-kilometer) radius were vaccinated, using day and night house-to-house vaccination. The area within a five-mile (eight-kilometer) radius was searched repeatedly as were the seven markets and nine schools. Locally hired launches were used to reach less-accessible areas, and a 500-*taka* ($33) reward was announced for anyone detecting a case of smallpox. No further cases were found.

In January 1976 a formal announcement was made that smallpox transmission had been interrupted throughout Asia. Two years of surveillance would be needed to confirm this with the certainty required for certification. It would take an unusual level of dedication and patience to sustain the activities of program staff and government leaders, especially in Bangladesh, if the world were to be persuaded that smallpox was no longer a threat.

Chapter 8

ETHIOPIA AND SOMALIA
The Last Countries with Smallpox

In the autumn of 1975, with smallpox conquered in Asia, our full attention shifted to the only remaining endemic country: Ethiopia, a country
of 30 million people (see figure 60). The beginning of the program there
had been delayed until 1971. Over the subsequent four years, a small but
highly motivated Ethiopian and WHO staff struggled heroically. They
were joined by young volunteers from the United States, Japan, and Austria, but despite their efforts, smallpox persisted tenaciously. Additional
manpower and resources were sorely needed. But other problems arose—
unrest and civil strife made many parts of the country unsafe, and some
areas became inaccessible to the teams. The atmosphere was volatile. We
were deeply concerned, and rightly so. Serious crises loomed.

Ethiopia is a far larger country than many realize: twice as large as
Texas, it is about the size of Germany and France combined. In 1971 it had
a health care staff of less than 3,000, including 350 physicians. The government's health budget was almost wholly devoted to curative care, not
prevention, but even then, fewer than 5 percent of the population had
access to medical care. In startling contrast was a large and costly mosquito
control program designed to eradicate malaria. Primarily supported by US
assistance, it was totally separate from the health services. It employed
8,000 workers scattered over one-third of the country.

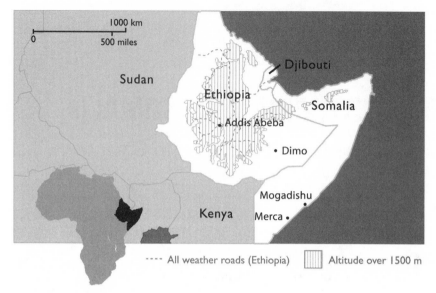

Figure 60. Ethiopia population (1970): 29,900,000.
   Somalia population (1970): 3,600,000.

The smallpox staff—until 1975, fewer than one hundred people—largely depended on mules and walking for much of the fieldwork. They continually had to contend with armed rebel groups and a feudal government structure. By late 1975, greatly strengthened by additional resources and veterans from other African and Asian programs, the staff began to stop transmission in area after area. We dared to feel hopeful. But in September 1976—just as the ultimate goal of global eradication seemed to be in sight—there came a startling report of smallpox cases from neighboring Somalia. Until then Somalia had been smallpox-free—or so we believed. Investigations soon turned up other cases, but the sources of infection could not be identified. Eventually it was discovered that a secretive and uncooperative Somali government was suppressing information about smallpox—in fact, the disease was widespread. Even greater difficulties arose, as frantic efforts were made to stop transmission before the disease once again escaped. We remembered all too well that this had happened before with serious consequences—when South African cases infected Botswana, and when infected refugees in India returned to Bangladesh. The outcome remained in doubt for many months.

SURVEILLANCE IN ETHIOPIA

Large mountainous areas of Ethiopia, inaccessible except by foot or on the back of a mule, made surveillance an unparalleled challenge. How could proper surveillance be done in a country of 30 million people with few health services, an unsophisticated political structure, and a smallpox staff of fewer than one hundred? A validation of the surveillance program was undertaken in June 1972 by a US Peace Corps volunteer, James Siemon, together with a local vaccinator, during a fourteen-day trek. At that time, several of Ethiopia's provinces appeared to have become free of smallpox, but were they? Siemon selected a district from which no cases had been reported for more than six months to determine whether it was truly smallpox-free.

The district covered 18,500 square miles (48,000 square kilometers) and had a population of 275,000. There was a health center and four clinics, eight schools (and 3,000 students), nine weekly markets, and eight Coptic churches. Using a homemade sketch map, Siemon and his colleague began with a visit to the district governor and procured a letter from him asking subordinates to provide food and accommodation. They then visited each of the health centers and schools and met with village leaders and shop owners. They carried with them the WHO smallpox recognition card and asked everywhere if anyone knew of a patient with a rash like that shown on the card. They timed their visits to reach the churches and weekly markets on those days when people would come from distances of seven miles or more.

Only one outbreak, of nine cases, was discovered, and vaccination containment measures were quickly taken. This outbreak had begun eight weeks before. The first case had come from a neighboring province. Seven of the next eight cases resulted from variolation.

Communication among residents in the district was surprisingly good, making the detection of cases more likely than had been foreseen. The team learned about the outbreak at five different locations (two schools and three markets), all between three and nine hours' walk from the outbreak location. Twenty other people with skin rashes were reported, but all turned out to have either chicken pox or skin infections.

ETHIOPIA

Ethiopia was one of the poorest African countries. Most of its population was widely distributed in small groups of huts scattered across the central highlands more than 5,000 feet (1,500 meters) above sea level. Rugged mountains and deep ravines throughout this area made travel extremely difficult most of the time and all but impossible during the rainy season, from June to September. There were lowland areas in the west and south-west with a fertile and more populated savanna grassland. Nomads roamed the vast Danakil and Ogaden deserts to the east and southeast, moving freely across the unmarked borders with Somalia and Djibouti (until 1977, called the French Territory of the Afars and Issas). Less than 7,500 miles (12,000 kilometers) of poorly maintained all-weather roads connected the few scattered, sparsely inhabited cities and towns. They accounted for less than 5 percent of the population. There were few health care units, and vaccination was all but unknown. Ethiopia had ten different ethnic groups speaking seventy different dialects. In rural areas where there were no roads, people simply walked—some for many miles—to markets, schools, and churches.

Our experience in several African countries and in Pakistan had shown that if smallpox transmission could be stopped in urban centers, the disease was likely to die out in the sparsely populated rural areas. Accordingly, at first, the strategy emphasized large-scale vaccination in urban areas plus surveillance-containment throughout the rural countryside. I believed that smallpox could not sustain itself for long in the thinly settled rural areas under these circumstances. But I was wrong.

Troubles in beginning the program

Although the program did not start until 1971, we had tried to begin its operations on a small scale more than two years earlier, but the government would not allow it. There were two reasons. The first was the fact that in Ethiopia, only the mild *Variola minor* was present—the form of smallpox similar to that in South Africa and Brazil. The disease was little more serious than chicken pox, and only 1 to 2 percent of patients died from it. Ethiopians did not fear the disease the way others had dreaded *Variola major*. The second reason was that the country was conducting a malaria eradication campaign and was convinced it could not undertake a second eradication program. At the time, the Ethiopian malaria eradication program was totally supported by the US Agency for International Development (USAID); the advisory staff were all from the United States.

I explained to the Ethiopian representatives at the World Health Assembly that eradicating smallpox did not require the thousands of workers it took to eradicate malaria. I proposed instead a small pilot program—to be established at WHO's expense—to assess the magnitude of the problem and to control any outbreaks they found. I offered repeatedly to go to Addis Abeba, the capital, to meet with the minister and other health officials, but permission to visit Ethiopia was always denied.

In 1968 Dr. George Lythcott, the senior US regional adviser to the West Africa smallpox program, was invited to go to Addis Abeba to address a meeting of the Organization of African States. I hoped that during this visit, he might have the opportunity to meet with officials at the health ministry and to discuss the importance of smallpox eradication. He was enthusiastic about doing so. However, shortly before his departure, US officials abruptly canceled his trip, never explaining why. We suspected, rightly, so we learned, that the malaria program advisers were behind the cancellation.

In September 1969 I sent yet another telex to the ministry of health proposing that I visit Ethiopia in conjunction with a trip I was planning to Kenya. To my amazement, the visit was approved. Permission was granted for a two-week visit for purposes of developing a national plan. I was accompanied by Dr. Ehsan Shafa, the WHO smallpox adviser from WHO's Eastern Mediterranean Regional Office (EMRO). The government's sudden change in attitude seemed too good to be true—and it was. We were not welcome after all. The minister informed us that he was not pleased to see us, that the government's position had not changed, and that the approval for our visit had been mistakenly sent by a subordinate in the minister's absence. He agreed to review whatever proposal we might have to make, but it was clear what his response would be.

We had few WHO resources that we could offer. We contacted a number of potential donors about making special contributions, but to little avail. Japan and the United States said that they might be able to make some volunteers available, but no other help was offered. The huge malaria eradication program, with its thousands of workers and hundreds of vehicles, seemed like it might be a helpful partner—but the answer was a vehement no. The program director flatly refused participation of any kind—even the loan of two or three vehicles for a few weeks. Years later, I learned why. The USAID senior malaria adviser said that he personally was responsible for Ethiopia's intransigence because he wanted no possible impediment to the success of the US-funded malaria program. Other priorities meant nothing to him, including any program accepted by WHO's World Health Assembly. (Even then, the Ethiopian malaria program was foundering. It soon collapsed.)

The plan we developed called for a WHO budget of $250,000, the provision of thirty Ethiopian health officers and assistants plus two WHO advisers, fifteen US Peace Corps volunteers, and seventeen Land Rovers. Resources would initially be concentrated in Addis Abeba and four of the fourteen provinces. With roads so bad and communications so difficult, this was the most practical approach we could conceive to provide suitable direction of the program. Additionally, we proposed that one sanitarian be assigned to each of the other provinces to begin building a reporting network. These workers would have to travel by bus and on foot. WHO funds would cover all costs except for the salaries of the Ethiopian health officers and sanitarians, who, we were told, could be transferred from other duties.

I presented the plan to the minister on a Saturday and was promised his answer on Monday. From his attitude, I knew what it would be. At a reception that Saturday, I met a tall, burly Austrian physician, Dr. Kurt Weithaler, who had served for the previous ten years as director of the Emperor's Imperial Guard Hospital. He was interested in the program and offered to serve as Ethiopia's principal adviser if the program were approved. Weithaler was also the personal physician to the imperial household. He said he might have the opportunity to speak with the emperor. On Monday morning a beaming minister of health told us how pleased he was about the projected program and signed the agreement.

The government asked that Weithaler be appointed "the responsible executive authority" as there were few trained or experienced Ethiopian staff at that time. To direct field operations, I was able to recruit Dr. Ciro de Quadros (see figure 61), who had performed so capably in the Brazilian program but was then leaving it. Ato Tamiru Debeya, one of Ethiopia's most capable sanitarians, was appointed to work with him. This central group served throughout the program.

Figure 61. Dr. Ciro de Quadros, a Brazilian epidemiologist, was Ethiopian director of field operations, astutely deploying limited resources and responsible for many imaginatively constructed epidemiological field studies. *Photograph courtesy of C. Quadros.*

The tip of the iceberg—1971–1972

When the teams prepared to begin field operations, I wondered just how extensive smallpox actually was. Only 197 cases had been reported in 1969 and 722 in 1970. The disease being the mild *Variola minor*, it was likely that cases were underreported, but I also believed that it would not spread readily in this sparsely populated country and might not be a serious problem. Our evidence was based, in part, on information that Shafa and I had acquired during our 1969 visit. We had questioned a great many nationals and internationals from different areas. Most of them reported that the severe form of smallpox had been present many years before, but only a few reported having heard of any smallpox cases at all in recent years.

The teams began work in January 1971 and found 275 cases in the first two weeks, 1,500 cases in February, and 3,400 cases in March. This amounted to half of all cases then being reported throughout the world.

About this time, I received a report from a recently arrived US Peace Corps volunteer, Vince Radke. He and his team were searching for cases by visiting health centers, village leaders, and schools. He said: "In the first classroom I visited, I obtained so many reports of cases in so many different villages that I went immediately to start investigating. Village after village was so heavily infected throughout large areas that we began to try to define the outer limits of the spreading epidemic, and to vaccinate people around the circumference of the area much as one would fight a forest fire."

Because of Radke's report, I cabled de Quadros to caution the staff about reporting outbreaks based on rumored cases without proper confirmation. I received an indignant reply: every case, de Quadros stated, was individually documented on a special form similar to one he had used in Brazil. (The Ethiopian program proved to be one of the most carefully documented of all the global programs.) By the end of the year, a field staff of only thirty-nine people had recorded and contained outbreaks with a total of 26,000 cases (see figure 62). Meanwhile, during the summer rains, the staff, using secondary-school students, vaccinated 154,000 people in Addis in an effort to stop urban spread. Neither this nor the heavy summer rains served to dampen the epidemic, however.

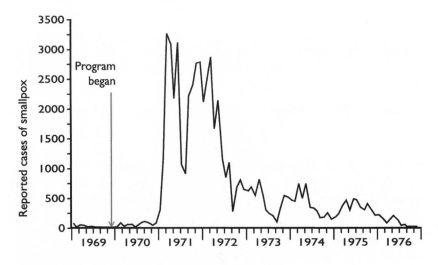

Figure 62. Ethiopia: Reported Cases (1971–1975).

More than half of the cases were from the four initially targeted provinces, where the teams had been working. What would similar work in the other ten provinces uncover? When an outbreak of smallpox in Kenya was traced to Ethiopia, all the smallpox-free neighboring countries expressed their concern to Ethiopian officials about its commitment to the global effort.

It was time to call in more help. I recruited two more WHO advisers, one of whom was Dr. Petrus Koswara, former director of the Indonesian smallpox program. Japan provided transceivers, vehicles, and twelve volunteers to serve as radio and automobile mechanics. Austria assigned four volunteers to the program, and the United States agreed to send Peace Corps volunteers as well. Cooperative help was sought wherever it could be found—including workers from an internationally sponsored leprosy-control program, mission groups, scientific expedition staff, emergency relief workers, and miscellaneous visiting health professionals from a number of different countries. They willingly worked long hours with no more than reimbursement for travel expenses. However, from the malaria program, no help was ever forthcoming.

It was soon apparent that there were four major problems: first, many patients, even when ill, traveled long distances on foot, transmitting disease wherever they went. Second, the ancient technique of variolation, whose practitioners took scab material and infected others by scratching it into the skin, was as widespread as in Afghanistan. In 1971 it was responsible for 12 percent of the cases. Third, the few available maps of the country were hopelessly incomplete and often wrong, making it necessary for teams to draw their own maps so they could identify the location of outbreaks for follow-up.

Finally, a great many people in central Ethiopia were opposed to being vaccinated. Throughout the global program most people had willingly accepted vaccination. But in Ethiopia there were several million Amhara and Dorsey people who rejected vaccination. Priests, village leaders, teachers, health workers, and others could not persuade them. Eventually, two approaches were found that proved effective: providing antimalaria drugs—only after vaccination—and administering vaccine with a jet injector. What there was about the jet injectors that made vaccination more acceptable was never clear.

Gradually, the number of staff and vehicles increased and the surveillance-containment operations extended into more provinces. The reporting was increasingly complete—but, despite better reporting, the number of cases fell steadily from 26,000 in 1971 to 5,400 in 1973. By September 1973 five of the fourteen provinces had become smallpox-free. All of Ethiopia's neighboring countries were smallpox-free, so there was little risk of importations.

At the end of the summer rains, we convened a seminar in Addis for Ethiopia and the neighboring countries to plan for what we hoped might be the last smallpox season. With willing assistance provided by Sudan, Kenya, and Djibouti, special search programs were undertaken in the difficult Ethiopian districts adjacent to the borders of each of the countries. We had high expectations of early success. Some US volunteers opted to extend their tours of duty and some returned to Ethiopia to help.

A SEARCH IN WESTERN ETHIOPIA: BAD ROADS, MACHETES, SNAKES

Searches were not easily undertaken in many parts of Ethiopia, but were especially difficult in Gojam Province, located in western Ethiopia adjacent to Sudan. With Sudan smallpox-free and anxious to be of help to Ethiopia, the governments agreed to have a Sudanese team cross the border to begin a combined search and vaccination effort across a 150-mile distance (240 kilometers) extending from the Sudanese border to the Ethiopian provincial capital of Bahir Dar.

Mr. Abdul Gadir El Sid, one of Sudan's most capable sanitarians, set off with twelve health workers, traveling in three Land Rovers and carrying with them fuel and most of the supplies they would need. It took them two months to traverse "roads" that had not been used for years. Bridges had to be built, and, in many areas, team members had to walk ahead of the vehicles using machetes to cut a path. On several occasions the underbrush had to be burned. One time, the flames nearly consumed one of the vehicles. Mechanical breakdowns, poisonous snakes, wild animals, and insects were daily annoyances. For the first part of the journey, the team could use Sudanese pounds to buy food; later, they needed Ethiopian dollars, which they raised by selling a supply of blankets brought along for this purpose.

They found no smallpox cases and vaccinated about 20,000 people en route. In Bahir Dar they were warmly welcomed by Ethiopian staff and provided with Ethiopian dollars. They returned home by the same route.

I left the September meeting in Addis to fly to New Delhi for a special meeting. India was anticipating the end of smallpox transmission within one year, by June 1974. It was launching an autumn drive with regular village-by-village and house-by-house searches for cases. Optimism prevailed in Bangladesh, as well, where there appeared to be no more than 150 infected villages in the entire country.

Disaster—drought, famine, hordes of refugees—1974

As the new year began, the tragic results of an extremely severe drought in northeastern Ethiopia were becoming apparent. Famine was widespread. As many as 200,000 people were believed to have died. Tens of thousands of refugees moved west into still-endemic central highland areas of Ethiopia where resistance to vaccination was strong and vaccination coverage was less than 20 percent.

The entire country was in turmoil. Open rebellion developed against the traditional feudal government, and sometimes entire districts were declared to be too dangerous for work to continue. Many of the teams traveled with a security escort. More resources were needed, but it was a difficult time to find such resources—the programs in Bangladesh and northern India were both facing major crises. Any discretionary funds that we could muster had to be directed to Asia because it harbored the severe *Variola major*.

A small staff in a very large country with a very limited road system posed serious logistics problems. Getting staff from one place to another remained a major hurdle (see figure 63). At the May World Health Assembly, I talked with US surgeon general Jesse Steinfeld about our transportation problems. His response was immediate, unexpected, and welcome. He arranged to transfer immediately $1 million to WHO, specifically to lease three helicopters for use in Ethiopia. Within three months a lease had been issued to a Canadian company and the helicopters were on their way. The helicopters were small (Hughes 500D)—able to carry only three passengers in the mountainous altitudes—and had a limited flight range, but they were priceless (see figure 64).

The arrival of the helicopters brought new challenges and another series of adventures. One helicopter had to make a forced landing when a fuel line was hit by rifle fire, and another was destroyed by a hand grenade. We were told that this was not meant as an attack on the program. It was

Figure 63. Ethiopian Bridges. Half the Ethiopian population was said to live more than two day's walk from any accessible road. "Accessible" was generously defined, and many bridges were all but impassible. *Photograph courtesy of WHO.*

because some villagers thought our program staff were Italians who were returning—the Italian occupiers having left thirty years before. One helicopter with its pilot was captured by rebels and a note was delivered to the government demanding that a large ransom be paid. Eventually, this was settled for a modest sum, and the helicopter and pilot were released unharmed. In the interim before his release, the Canadian pilot persuaded the rebels that they all should be vaccinated, and this he did himself. As time went on, most of the Canadian pilots became as knowledgeable about smallpox and the program as any of our senior staff.

Active fighting extended across the country. A long-standing revolution in Eritrea intensified; violence grew among the Amharic peoples in the central highlands; and Somalis throughout the Ogaden desert began to engage in sporadic open warfare, prompted by forces called the Western Somali Liberation Front. Some of those active in hostilities arranged for the smallpox staff to work in areas they controlled. Others attacked the teams and kidnapped them for varying periods. All were usually released unharmed within a week or two.

Figure 64. Helicopter. With a large country, few staff, and roads in terrible condition, three small helicopters proved to be a critical addition. *Photograph courtesy of WHO.*

In September 1974, Ethiopia's emperor was deposed and a revolutionary government installed. Somehow de Quadros and Ato Tamiru managed to keep the program going, and by the end of the year eight of the fourteen provinces were smallpox-free. During 1974, a total of 4,400 cases were discovered—a decrease of 20 percent from the preceding year. Considering the problems, however, this was a remarkable achievement.

Floods, fighting, and unexpected news—1975

The civil unrest increased throughout the first half of 1975, but the tempo of the program was sustained, albeit with mounting difficulty. In July, de Quadros began to define progress as they did in Asia, by the numbers of active outbreaks. He determined that there were 131, almost all confined to three provinces, in central and north-central mountainous districts, some of which were considered unsafe to enter. Civil disorder had become so widespread and dangerous that several countries pulled back all personnel from the countryside and evacuated some from the country entirely. Volunteers from Austria and the United States were withdrawn,

THE HAZARDS OF AIR TRAVEL IN ETHIOPIA

There are few places that rival the sheer beauty of mountainous Ethiopia—as well as few places where, day to day, the staff encountered more potentially threatening personal hazards and perils. Two of my own experiences are illustrative.

One adventure involved running out of gas over the Blue Nile Gorge in a single-wing two-passenger plane. The plane had been leased to permit me to meet with staff some two hours away from Addis—travel by road would have required days. On the return flight, the motor sputtered and died. We were directly over the gorge—5,000 feet (1,500 meters) above the Blue Nile—at least a mile from the edge of the canyon. The pilot was a distressingly casual Swede who mentioned that he had forgotten to put on the gas cap. The gasoline had drained out. He switched to the second tank and eventually managed to get the engine restarted. As we approached the Addis airport, the engine again began to sputter, and the pilot called the airport to announce an emergency landing. However, an Ethiopian passenger jet was descending at that time. The tower and the two pilots engaged in a heated argument before the jet finally pulled up. As we began to taxi, our plane's engine stopped altogether. Two months later the Swedish pilot was killed and the plane crashed when a large buzzard penetrated the cockpit window and impaled itself in the pilot's chest.

The second unwanted adventure involved myself, an Ethiopian health officer, and the pilot in a helicopter lost at night in a rebel-infested scrub desert area of southern Ethiopia. It was 1979, the time of final confirmation of eradication. We had checked the program in a small town—the next stop was another small town about sixty minutes away. Unexpectedly, a cloud bank moved in from the west, and nightfall came suddenly. It was pitch dark, there were no lights on the ground, and no radio navigation. The helicopter had lights to illuminate the ground, but, on scrub desert, there were stunted trees that endangered landing and dense clouds of sand that totally obscured visibility. The town was known to have only a few lights driven by a generator. Long after we expected to arrive, a single tiny light became visible; in fact, it was probably a 60-watt lamp at the top of a pole. As we got closer, small kerosene lights could be seen. A central area looked clear; we descended, hoping that the blades did not clip something. The plane was surrounded by men with submachine guns. Some fifteen minutes later, a smallpox vaccinator mysteriously appeared and explained who we were.

and except for smallpox program personnel, foreign nationals were forbidden from working outside Addis.

With the interruption of transmission in India and imminent interruption in Bangladesh, it became possible to transfer some veteran WHO advisers and resources to Ethiopia. More donations were received. Monetary contributions came from Canada, the Netherlands, and USAID.

That summer of 1975, we met again to map out plans for the next eighteen months. The revolutionary government agreed to issue a proclamation declaring the program to be of highest priority. The new plans called for increases in WHO advisers from 7 to 25 and Ethiopian staff from 42 to 125. This was in addition to 1,000 locally recruited workers. The program received an unexpected boost in manpower when the government decided to send 60,000 young secondary-school students into villages throughout the country in a chaotic campaign supposedly intended to increase literacy. Some of these students, the government agreed, could work in eradicating smallpox. The students provided invaluable help. As natives, they were better able to approach political leaders and villagers about reporting smallpox and accepting vaccination.

By the end of 1975 the staff was discovering fewer than ten outbreaks per week. The number of infected villages was down to sixty, and these were confined to just two areas. One was in and adjacent to the Blue Nile Gorge, a virtually impenetrable one-mile-deep gorge extending for more than two hundred miles, which was held by armed and hostile dissidents. The other was among nomads in the vast Ogaden desert.

Intensive surveillance-containment operations around the edge of the Blue Nile Gorge quickly limited any outbreaks that spread outside the gorge until eventually, with the use of civil guards, it was possible to penetrate the gorge itself and stop the last outbreaks.

Not many resources had been assigned to the Ogaden Desert since it was not foreseen to be a problem. The population of about 300,000 nomads was scattered. An estimated 60 percent of them were believed to be protected by vaccination that had been administered at oases and along a river. Because the nomads roamed freely across the border with Somalia, periodic meetings were arranged with the Somali smallpox program staff to coordinate activities. These had been reassuring. The Somalis reported that 85 percent of the desert nomads in Somalia were protected and that surveillance in Somalia was thorough.

During the early months of 1976, scattered outbreaks occurred among nomad groups in the Ogaden desert. But the sources of infection were seldom able to be traced. Taking into account population size, the widely dispersed nomadic groups, and the level of vaccination, it seemed unlikely that smallpox could persist for very long. It was a rational but ultimately costly and near fatal assumption.

Additional teams and advisers were assigned to the Ogaden desert area in March. However, operations were hampered by Somali guerillas, whose cooperation was unpredictable at best. Although these forces usually made an effort to protect the staff, on nine different occasions they seized vaccination teams and took them into Somalia. Each event was marked by considerable anxiety and uncertainty, but special interventions by United Nations officials secured their release. On one occasion the guerillas warned that an attack was planned on an outpost; the team called for a helicopter to evacuate them. As the helicopter took off, they saw the attack begin. On other occasions the guerilla forces attacked government buildings, killing the occupants but sparing the people working in the smallpox program.

By the summer of 1976 it became apparent that smallpox was still being spread in farming settlements along a river near the border. But once again nature intervened. The most serious floods in decades destroyed what few roads and river crossings existed and scattered the settlers. This border area also became the zone for serious fighting, making it impossible for teams to enter the area until mid-July.

On July 22 an outbreak was discovered in a small nomad village called Dimo. Investigation revealed that sixteen cases had occurred, nine resulting from variolation. The last to become ill, on August 9, was a three-year-old girl who recovered uneventfully. She had been variolated as well as vaccinated during the incubation period.

Searches continued throughout the Ogaden and in the rest of Ethiopia as well, but no further cases were found. Meanwhile, in the highlands, the last case had occurred on July 5. Week after week went by, and we prepared to announce to the world at the end of October that ten weeks had elapsed since the case of three-year-old Amina Salat—presumably the world's last case of smallpox.

I flew to Ethiopia to accompany a television crew taking pictures of the village and the last patient. The enormous size of the area and the lack of landmarks were startling. We had to circle over the general area of the outbreak for nearly half an hour before the pilot could find the village that he had visited several times during the preceding weeks.

On September 27 I had a call from the CDC reference laboratory. They had received two specimens from Somalia that contained smallpox virus. I cabled immediately to the capital, Mogadishu:

POXVIRUS PARTICLES PRESENT BOTH SPECIMENS STOP URGENT THAT EVERY CONTACT SINCE RASH ONSET BE FOUND AND VACCINATED INCLUDING ALL HOSPITAL PATIENTS STOP SITUATION MOST CRITICAL SINCE NO SMALLPOX SINCE AUGUST 9 IN ETHIOPIA STOP...HIDDEN FOCUS MUST BE PRESENT NEAR BORDER OR POSSIBLY ELSEWHERE STOP THIS COULD BE WORLDS LAST FOCUS STOP ESSENTIAL THIS BE FOUND AND CONTAINED URGENTLY STOP ARITA DEPARTING EARLIEST POSSIBLE FLIGHT TO ASSIST.

SOMALIA: AN EPIDEMIC THAT SHOULD NEVER HAVE HAPPENED

The report of smallpox in Somalia came after a year in which we had celebrated victory in India, Bangladesh, and most recently Ethiopia. It was a bitter disappointment for the smallpox staff and an embarrassment to the Somali government, which, for years, had sharply criticized the Ethiopian government for its deficiencies. WHO staff quickly flew to Somalia to assess the situation, but government officials curtailed their movements and provided little help. Soon it became apparent that reports of known cases were being suppressed. Far more serious problems loomed than any could then imagine.

When the global program had begun, we believed that Somalia was smallpox-free. However, because it shared a long, open border with endemic Ethiopia, it seemed prudent to develop a special program there. Shafa, the WHO regional smallpox adviser, working with the Somali government, drew up a conventional plan that provided for a three-year massvaccination program but said regrettably little about surveillance. In 1969, a WHO adviser was assigned by WHO's Eastern Mediterranean Regional Office (EMRO) to Somalia.

The Somali people, predominantly nomadic and seminomadic,

roamed across the open and unmarked borders in areas that included Somalia as well as parts of Ethiopia, Kenya, and Djibouti. For years the socialist military government hoped to unite all the Somali people in one nation. This led to continuing animosity between Somalia and its neighbors, especially Ethiopia.

In 1975 Somalia's population was 3.6 million. Most of the people were herdsmen, caring for an estimated 23 million goats, sheep, camels, and cattle. Basically, the country was divided into two major population areas: one in the north with about 1 million people, and one in the south with 1.5 million people. The remainder was spread over an intermediate 500-mile semiarid zone. Few people traveled between the northern and southern zones. Cases of smallpox eventually were found to have been restricted to the southern zone, centered on Mogadishu, the only large city (population, 450,000). Two rivers ran through the southern area providing irrigation for farming communities and a source of water for seminomads, who had large herds of cattle, sheep, and goats. During the rainy season, from March to June, they moved the herds away from the rivers toward southern Somalia and Kenya to escape the hordes of tsetse flies. In addition there were large numbers who were fully nomadic, with herds made up only of cattle and camels. During the rainy season they moved every two to three weeks, their destinations wholly unpredictable. Rain fell irregularly throughout the area, often in local downpours, and nomad groups of perhaps three or four families would quickly pack up all their belongings and move their herds as far as twenty to thirty miles away to take advantage of water and forage.

During the time of the eradication program, Somalia's population was especially mobile. It seemed at times as though most of the population was moving or had just migrated from one area to another. Large numbers of migrant laborers came north from throughout southern Somalia, as well as from Ethiopia, to work on the banana and sugar cane plantations. A severe drought throughout the Ogaden Desert in 1974 and 1975 gave rise to some 200,000 to 300,000 refugees who had to be housed in special camps. Another cause for migration was a program begun in 1975 to resettle more than 100,000 nomads in southern agricultural areas. Later, in 1977, tens of thousands of people in the border areas fled inland to escape the Somali-Ethiopian war (see figure 65).

On the plus side, Somalia's roads and health services were far better and more extensive than those in Ethiopia. For a population only one-tenth as large as Ethiopia's, there were more than 1,300 health personnel, 21 hospitals, and 187 health posts.

Smallpox before 1975

In 1962 Somalia had experienced an outbreak of 221 cases of smallpox. For the next thirteen years, it was smallpox-free except for a few imported cases from Ethiopia. Between 1962 and 1970, fewer than 50,000 people each year were vaccinated—and this was with a liquid vaccine of doubtful potency. Scar surveys in 1967 and 1968 showed that almost no children under age five and fewer than half of older children and adults had vaccination scars. Surveillance was poor. It is probable that many more smallpox cases had been imported than had been known or acknowledged. However, with a broadly dispersed population, significant outbreaks did not occur.

The WHO-supported mass-vaccination program had begun in August 1969, using freeze-dried vaccine. By the end of 1974 the teams reported

Figure 65. Refugee Camp, Somalia. Famine, floods, and war produced numerous refugees. Coupled with the large number of migrant workers and resettlement schemes, it seemed as if all Somalians were continually on the move. *Photograph courtesy UNHCR.*

having vaccinated 3.5 million people. Information obtained later suggests that coverage in the seminomadic and settled populations may have been as high as 80 percent. Among nomads in the south, it was not more than 20 percent. Little was done during this period to improve reporting.

Growing suspicions

Beginning in October 1972 the number of imported cases began to increase. By February 1976, thirty-eight cases had been reported. It was reported that smallpox had spread to others on only one occasion. I was suspicious. No other country—even one with a good surveillance system—had been this successful in rapidly detecting and containing imported cases. More puzzling was the fact that twenty of the cases were supposedly detected within three days after the onset of the rash. This was absurd. Experienced clinicians are seldom confident about a smallpox diagnosis until at least the fourth or fifth day of the rash. Our doubts were further heightened by the fact that some cases had first been reported to me in confidence by the Soviet embassy. However, they were not reported by the Somali government until we asked for confirmation. We repeatedly asked our WHO smallpox adviser in Somalia, Dr. Mahfuz Ali, from Pakistan, to provide further information, but we received only his reassurance that all was well.

In retrospect, we should have undertaken an independent on-site assessment, but we were totally absorbed in the crises in India and Bangladesh. We continued to believe that sparsely populated Somalia—which had so recently completed a mass-vaccination program—was an unlikely place for smallpox transmission to become reestablished, however inadequate the surveillance program. Moreover, a WHO smallpox adviser was on the job there. After February 1976, no further cases were reported from Somalia until September, from either the government or from the Soviet embassy.

Smallpox uncovered in Mogadishu—September 1976

We believed we had seen the world's last case of smallpox on August 9, 1976, in Dimo, Ethiopia. Our first indication that smallpox still persisted—but in Somalia—was the September 27 CDC report informing us of smallpox virus in specimens from two patients. Dr. Arita flew immediately to Mogadishu from Geneva and discovered there were not two but five patients. Each of them claimed to have walked from the Ethiopian Ogaden desert area to the Somali border and then traveled more than 300 miles (480 kilometers) by bus to Mogadishu. The inference was that they had been infected in Ethiopia. Arita alerted de Quadros in Ethiopia, who mounted a major search throughout an area of more than 25,000 square miles (65,000 square kilometers). Thirteen epidemiologists and 150 search workers with vehicles, helicopters, and fixed-wing aircraft participated. No cases or even rumors of cases could be found.

Where the five people had become infected remained a mystery. No less puzzling was why two specimens from smallpox cases were sent to the CDC for diagnosis when Somalia had never submitted specimens to any laboratory before this time. It was not until months later that we learned some of the ugly facts. The Ethiopian search had indeed been futile. By then, however, endemic smallpox had already become widely established in southern Somalia. This fact was known to Somali staff and our thoroughly untrustworthy WHO adviser. As we eventually learned, there had been other smallpox patients, hospitalized as early as June, and other outbreaks earlier in the year. The Somali government had hoped to hide these outbreaks and to avoid the publicity and embarrassment of being the last country in the world harboring smallpox.

Two events appear to have been responsible for Somalia's sudden decision to make the smallpox cases known. The first was a letter I had sent to WHO's regional offices with a copy to staff in Ethiopia, Kenya, Sudan, and Somalia:

> As the days go by, it seems increasingly possible that the Ethiopian case with onset on 9 August 1976 could be the world's last case of smallpox. …My guess would be that it will be difficult to have reasonable confidence about this until late October. When we reach that point, the Director-General would like to make a major announcement…conceivably jointly with the Secretary General.…(O)ne would not wish to make an announcement that there are no known smallpox foci without having reasonable confidence that another outbreak would not emerge one or two weeks later.

The second event was the unexpected arrival in Mogadishu of Dr. Bert van Ramshorst, a Dutch WHO Ethiopian smallpox adviser. Captured by Somali "border police" (actually working well inside Ethiopia), he was held for four days until a United Nations official in Mogadishu could arrange for his release. Meanwhile, he had arranged to have a meeting with the minister of health. He succeeded in persuading the minister that the Ethiopians—contrary to the Somali government's belief—were operating a highly effective program and that smallpox transmission had almost certainly been stopped. The minister demanded absolute confirmation that his Somali cases were actually smallpox. This was why two specimens had been sent to the CDC.

Where was the source? The patients provided no helpful information. Area-wide searches were conducted. Over the following month Somali staff and two WHO advisers contacted about 12,000 people in thirty different villages. Vaccination coverage was surprisingly good with 70 to 90 percent showing vaccination scars. All denied having seen cases of smallpox for two years or more. Some 2,200 Somali staff conducted two house-to-house searches in Mogadishu. Nothing was found (see figure 66).

And still, more new cases of smallpox—twenty by the end of October—continued to trickle into the hospital. Most of these could be linked to earlier cases, although this information came via Somali interviewers. The WHO epidemiologists were not permitted to question patients or their contacts.

During the first two weeks of November, a citywide mass-vaccination program was conducted, followed by a night search for cases. None were found. In frustration, the four WHO epidemiologists proposed an ambitious search over the five southern administrative regions. WHO Director-General Mahler approved the additional costs involved and sent a telegram to the minister of health. Mahler expressed his hope that at the January 1977 WHO executive board meeting, he would be able to announce that the world's last focus of smallpox had been eliminated. The minister replied: "[W]e shall spare no efforts to mobilize all available resources." However, permission for the search was not forthcoming until the Somali government received word that Assistant Director-General Ivan Ladnyi would visit personally in early November to review the situation. The minister then granted approval for WHO staff to make visits—of one day only—to each of the regions. He demanded that $2 million be

Figure 66. The Search for Cases. The WHO recognition cards were used extensively by search teams inquiring about possible cases of smallpox. *Photograph courtesy of WHO/J. Magee.*

made available to pay for the more extensive search. This was far in excess of what was required. The demand was rejected.

The number of patients hospitalized with smallpox decreased from nine in November to five in January. Then, suddenly, there were none. Staff were concerned that if searches were conducted only during the day, patients might be hidden or sent to other parts of the city. Accordingly, night searches were organized for the entire city in December and January. No cases were found.

By the time of the January 1977 executive board meeting, there were no known infected areas. However, without adequate searches, no one could be confident that the country was really smallpox-free. Approval for an organized national search was finally granted in January during a Geneva meeting involving Mahler, Ladnyi, our program staff, and the Somali representative to the executive board. Arita organized this search, which began in late March as a coordinated effort across Somalia, southern Kenya, and neighboring areas of Sudan. It was expected that this would be the first stage in the final confirmation of eradication. For the search, WHO agreed to provide an additional $350,000 and to add six more epidemiologists.

Lies, cover-ups, and secret records

The actively obstructive Somali government officials proved to be a problem as intractable as any the program had encountered. But in February, government cooperation suddenly began to improve. This occurred shortly after the appointment of a new national smallpox eradication program manager, Dr. Abdullahi Deria, who had just returned from studies at the London School of Tropical Medicine and Hygiene. He brought to government an understanding of the issues and persuaded a reluctant, suspicious minister to stop hindering the program.

The national search involved 300 Somali workers and four WHO advisers. Reports from outside Mogadishu began to flow in, and by mid-April, twenty-nine cases had been identified. There was yet more disturbing news: the teams discovered that an outbreak had occurred in November and that it had even been verified by WHO's own national adviser—but the report had been suppressed. Even more startling was the discovery that patients had been regularly admitted to two different parts of the Mogadishu hospital—one part was open to WHO staff and one part was kept secret. In the open ward, thirty-four cases were officially registered. In the secret area, more than 500 had been hospitalized!

The number of cases continued to increase, reaching 636 in May. Arita hurriedly flew from Geneva to Mogadishu and decided to recruit

---

A NOVEL WAY TO DETECT HIDDEN CASES

In March 1977 new instructions from the Somali ministry of health urged that cases now be reported, but the old policy of concealment continued to be widespread. One of the first to break the barrier was the redoubtable Sudanese sanitarian, Abdul Gadir El Sid, who was now a WHO adviser. He had been the leader of the Sudanese surveillance team described earlier in this chapter. On entering a village for the purpose of investigating suspected cases, he saw several people with facial pockmarks suggestive of recent smallpox. But all of his questions came back with unanimous denial by the villagers that there were cases. Taking over the vehicle from his driver, he deliberately drove it into deep mud. A large crowd came from the village to help extricate the car. Among them were four people with active smallpox.

Figure 67. Isolation Hut. A hut in which a nomad with smallpox is isolated. The huts were built away from the encampments and enclosed by a thorn-bush barrier to keep out wild animals and visitors. A containment team member makes certain that everyone entering the enclosure has been recently vaccinated. *Photograph courtesy of T. S. Jones.*

another eight WHO epidemiologists. It was the first step toward a greatly intensified effort, which, at its peak in June and July, included forty-four WHO and national epidemiologists, 2,700 surveillance workers and supervisors, and nearly 600 watch guards to make sure that patients remained isolated (see figure 67).

Delegates attending the May World Health Assembly expressed concern about the burgeoning epidemic and the potential for spread by pilgrims in November when the annual Haj pilgrimage to Mecca began. With Somalia's proximity to Saudi Arabia, many Somalis made the overland trek by bus, by camel, and on foot. There was reason to fear that some patients, having only mild symptoms, would join the pilgrimage. If the disease spread among the hundreds of thousands of pilgrims in Mecca, smallpox could be disseminated across the world.

Once again there was the ubiquitous problem of a shortage of resources—a crisis similar to those we had faced over the preceding two years, first in India, then in Bangladesh, and finally in Ethiopia. This time, however, Somalia declared the smallpox epidemic a disaster, and the United Nations Disaster Relief Operations agreed. A general appeal for assistance brought special donations from Canada, the Netherlands, Norway, Sweden, the United Kingdom, and the League of Red Cross Societies. CDC director Dr. William Foege sent five CDC epidemiologists on emergency assignment; France provided three teams and vehicles; OXFAM, the British charity, sent four more teams. All supplies, including vehicles, came by special airlifts arranged by Sweden, Canada, and the United Kingdom. Somali political party secretaries worked to mobilize party workers. Other local groups offered help—the Somali Women's Democratic Organization, the Somali Workers' Organization, and the Somali Youth League.

The numbers of cases rose to 1,388 in June and the number of infected villages to 223. Containing the outbreaks, especially among the nomad populations, proved to be a challenge quite unlike any faced before. The proportion of outbreaks among nomads rose from 50 percent in April to 75 percent in June, and more than 90 percent in July.

The nomads typically began their most active migrations in March when the rainy season began. They tended not to follow established patterns of movement but to seek good grazing for their animals wherever that might take them. Their vaccination levels, wherever checked, were usually in the range of 10 to 20 percent. Simply finding the nomads in the scrub desert was an adventure. Line-of-sight vision was blocked in every direction by the high scrub. No maps were available and few landmarks existed. The only good lead was the knowledge that the nomads must find water holes. In some areas, lookouts were posted at every water hole to keep a vigil for the nomads.

A surprising discovery was that smallpox could persist for long periods in small nomad groups. In one group of forty-four people—only seven of whom had ever been vaccinated—nineteen cases occurred over a five-month period while the group roamed over an area of nearly 250 square miles (650 square kilometers). In other parts of the world, infection in a family group such as this usually did not last more than six to eight weeks before dying out. Here, the difference was the less infectious *Variola minor* and the fact that a nomad family was not confined to just a few small huts.

The smallpox-affected localities steadily diminished in number—

from a high of 220 in June to only 29 in September. But then torrential rains hampered the use of heavy vehicles. Many field staff were obliged to travel on foot or on camels and donkeys. The teams found only five outbreaks in October, all among nomads. On October 31, 1977, a final case was discovered in the port town of Merca.

THE LAST CASE

Ali Maow Maalin was a twenty-three-year-old cook at the local hospital. He developed fever on October 22 followed by a rash on October 26 (see plate 5). His case was a classic one in depicting omissions and mistakes in program operations. He had never been vaccinated despite having once served as a vaccinator and despite having worked at the hospital where employee vaccinations were supposed to be mandatory. On October 12 two sick children arrived at the hospital in a vehicle from a nomad encampment. They were to be housed in an isolation camp nearby. Both of them had smallpox, and one died two days later. Maalin volunteered to ride with them to direct the driver to the camp about 200 yards away. His exposure was brief but adequate.

Maalin was admitted to the hospital on October 25 with a presumptive diagnosis of malaria. He received numerous visitors and walked freely around the hospital and outside the compound. A day later he developed a rash that was diagnosed as chicken pox and he was sent home. A popular man, he received many visitors until October 30 when a male nurse suspected that Maalin had smallpox. He was then sent immediately to the isolation camp.

An intensive search began to find everyone with whom he had come into contact. In all, ninety-one face-to-face contacts were identified, twelve of whom had no vaccination scar, and six who had been hospital patients or visitors. Heroic measures were taken, including a search and vaccination of the town and of everyone entering or leaving town at any one of four checkpoints. House-by-house searches throughout the region were conducted monthly, and a national search was completed on December 29.

Between the declaration of a state of emergency and the last case, only 141 days had elapsed. Veterans from around the world had surmounted one more unprecedented challenge. The epidemic was stopped before the annual pilgrimage to Mecca got under way. Meanwhile 3,022 cases had occurred.

Ali Maow Maalin survived his illness and continued to reside in Merca doing a variety of different tasks. But he has a place in history as the last naturally occurring case in a continuing chain of transmission extending back at least 3,500 years.

The work was not yet finished, however. There was "only" the need to ensure that sufficiently sensitive surveillance programs were in place around the globe for at least two years in order to satisfy ourselves and health officials throughout the world that smallpox had truly been eradicated, that there was no remaining focus of the disease anywhere, and that vaccination could cease.

Understandably, health officials in every country were anxious to begin utilizing smallpox staff to work in other programs. Persuading them, and all smallpox eradication staff, of the critical need to continue activities was a task almost as formidable as stopping smallpox in the first place. Fortunately, many of the veteran, dedicated staff stayed the course.

Chapter 9
SMALLPOX—POST-ERADICATION

The eradication of smallpox meant that vaccination could be stopped everywhere; that a vaccine that had been in use for more than 175 years was no longer required; that there was no longer a need to fear a recurrence of the dreaded smallpox, which had ravaged the world for more than 3,500 years. A daunting task remained, however. We ourselves had to be convinced there was no remaining smallpox focus anywhere in the world. And we had to have evidence of this sufficient to persuade national authorities and the public. Only then would the threat and the fear be lifted.

The challenge was unique, because never before had a disease been eradicated worldwide. Skepticism prevailed among those who had lived or traveled in endemic countries. They knew personally the vastness of Asia and Africa; the lack of roads, health services, and communications in many areas; and the inevitable disruptions of programs by civil wars, floods, famines, and refugees. I confess to having had doubts myself that we could know with certainty that no cases existed. I was reminded of this on each occasion I thought of Calcutta, India, in particular. How could any program of activities ensure that there were no cases whatsoever among the teeming masses of people extending for miles throughout such a vast city? Meanwhile, there were academics who pointed out that we live in an ecologically interconnected world—that it was technically impossible to deliberately eliminate one organism from this complex milieu.

Thus, the containment of the last-known smallpox outbreak concluded the opening chapters of the eradication saga, but at least two more years would be required to confirm eradication—and these were difficult years. The smallpox staff no longer had the impetus of epidemic smallpox or the threat of its imminent importation. National health directors were in need of more personnel and vehicles for other programs. They were persuaded with difficulty to undertake needed surveillance programs to prove that no smallpox cases existed.

Sustaining the motivation of program staff once the last cases had been contained was destined to be difficult. The importance of completing the task was frequently enjoined in the simple phrase "the final inch,"

"REALM OF THE FINAL INCH"

This message to the staff was sent in Progress Report 32, October 27, 1975:

"We are entering a most critical phase where optimism and/or complacency could prevent us from reaching our ultimate goal. Professor Holger Lundbeck, Swedish National Bacteriological Laboratory, has called our attention to Solzhenitsyn's 'The Final Inch' from the *First Circle*" (Lundbeck's translation).

> The realm of the Final Inch! . . . The work has been almost completed, the goal almost attained, everything seems completely right and the difficulties overcome. But the quality of the thing is not quite right. Finishing touches are needed. . . . In that moment of fatigue and self-satisfaction, it is especially tempting to leave the work without having attained the apex of quality. . . . In fact, the rule of the Final Inch consists in this: not to shirk this crucial work. Not to postpone it. . . . And not to mind the time spent on it, knowing that one's purpose lies not in completing things faster but in the attainment of perfection.

making it clear that the program was not completed until the ultimate finish line of the race had been crossed.

In February 1977 global leadership of the program was assumed by Dr. Isao Arita, my deputy since the program began. He was ably assisted by a remaining cadre of program veterans. Reluctantly I left Geneva to become dean of the Johns Hopkins School of Public Health, but I was able to continue part-time participation and to follow the program closely, albeit at a distance.

ERADICATION—WHAT DOES IT MEAN AND HOW DO WE DEFINE IT?

From the earliest days of the global planning process, there were two different views as to what was meant by eradication. A number of health professionals insisted that it had to mean the extermination of the virus from the planet. To me this was unrealistic. All the virus needed to start an epidemic could be retained indefinitely in one tiny vial in a deep freezer, perhaps forgotten even by the virologist who had isolated it. We had no way to verify that no vials of smallpox remained in any laboratory anywhere in the world.

The only reasonable approach was to define eradication in terms of the absence of human cases. If there were no human cases, there could be no circulating virus. We were reasonably confident that animals could not become infected and that there was no natural reservoir in nature. Epidemiological studies during the program supported these beliefs. For the virus to sustain itself, it had to pass from one person to another in a continuing chain. Each victim would develop the characteristic disease and each would either die or, after recovery, be fully immune for the rest of his life. Any remaining focus of smallpox had to involve an ever-spreading chain of infected people.

Confirming eradication, therefore, required the development of systems to detect cases in the infection chain. No system could be perfect enough to detect every case, but the probability of detecting at least one case in a chain steadily increased over time. How long must a country or continent sustain a competent surveillance system before it could have confidence that no remaining focus existed? Initially we postulated an interval of two years. From national program experience, we had found that the longest interval during which a virus was circulating in a country, unreported, was eight months. Our prescribed two-year interval was three times the eight-month period and so this standard was used. The one eight-month delay occurred in Indonesia because local health officials suppressed reports out of fear they might be punished for insufficient effort in promoting vaccination. To prevent this from occurring again, a reward for anyone reporting a case of smallpox was widely publicized (see plate 6). Large numbers of people with suspicious rashes were brought forward, including occasional unhappy tourists with odd sorts of pimples.

SURVEILLANCE AND SEARCH

Establishing ongoing programs adequate to find human cases required imagination and resourcefulness. The results had to be sufficient to convince an international commission, which would review the program in each country at least two years after the last case. Surveillance was an essential component of the eradication strategy, and thus routine reporting systems were promoted in all countries. In addition, all endemic countries and those at heightened risk undertook special search activities to detect possible cases. To assist countries during the two-year confirmation period, WHO frequently assigned short-term consultants for periods of weeks to months.

A routine reporting system

A network in which all health centers and hospitals reported cases every week was a fundamental need. Most countries were able to achieve the goal of having 80 percent of all reporting sites providing information within two weeks—even in areas where reports had to be hand-carried by messenger. When smallpox cases stopped occurring, many countries began asking that chickenpox cases be reported. It provided a separate check to keep centers alert.

Special searches

Every program supplemented the routine reporting system with systematic searches of special areas thought to be at high risk. Most of the searches were "scar surveys." Individuals were checked for vaccination scars to determine the completeness of vaccination coverage and to detect facial scars from smallpox if present. As we discovered, 80 percent of *Variola major* victims bore permanent facial scars. Particular attention was directed to young children to see if any had acquired scars since the last known cases had occurred. Areas considered at highest risk were those where reporting records were poorest, where migrant groups congregated, where poverty was greatest, and where smallpox cases were known to have occurred during the years of the program.

Rumor registries

All endemic and at-risk countries were asked to establish "rumor registries"—effectively, to document all cases reported as suspect smallpox and the outcome of each investigation. The number of such cases provided an indicator as to the sensitivity of the system. Specimens from suspected cases were regularly submitted and processed by one of the two WHO international collaborating laboratories—the CDC in Atlanta and the Research Institute for Viral Preparations in Moscow. Between 1978 and 1980, 8,930 specimens were examined—none contained smallpox virus.

INTERNATIONAL COMMISSIONS

To enhance acceptance of a claim that eradication had been achieved, WHO appointed independent international commissions to certify smallpox eradication in each country or small group of countries that had been recently infected or were at special risk. Precedence for a certification procedure had been established during the malaria eradication program and was readily accepted by the countries.

We sought to appoint individuals to the commissions who had doubts about the success of the eradication program, either generally or regarding the particular country concerned. The pattern was for the commission to meet at the capital city for two to three days to review the program and the data and to proceed to the field for one to two weeks as individuals or pairs to visit areas of greatest concern. The commission reconvened at the end of its visit, at which time it was asked to reach one of two conclusions: either that it was satisfied that eradication had been achieved or that it would be satisfied if additional specified studies were conducted.

As noted earlier, the first of the international commissions—for South America—was convened in 1973. It was poorly conducted and of limited validity. Fortunately, subsequent events confirmed that smallpox had been eradicated. A more rigorous protocol was devised for the second international commission, in Indonesia, and this served as a template.

Between 1973 and 1979, thirty-six countries were visited by twenty-one different independent commissions. Special reviews of programs were performed in seventy-nine countries and eradication was certified in all.

GLOBAL CERTIFICATION OF SMALLPOX ERADICATION

As the work of the independent commissions was concluding, we perceived the need to establish an overall WHO Global Commission for the Certification of Smallpox Eradication. It would review all of the work done in confirming eradication. The commission would be asked to recommend any additional activities it deemed necessary in order to have full assurance that eradication had been achieved worldwide. Twenty-one public health and scientific experts from nineteen different countries were appointed to the commission. They held two meetings—in 1978 and 1979.

The chairman of the global commission was Dr. Frank Fenner, acknowledged to be the world's foremost expert in the virology of smallpox and related viruses. He was professor and former director of the John Curtin School of Medical Research at the Australian National University. Fenner had worked closely with the smallpox eradication program since 1969, having chaired our committee on research and having served on international commissions for certification of India and Nepal and the countries of eastern and southern Africa. One of Australia's foremost scientists, he had numerous national responsibilities but was ever willing to offer whatever we requested in time, advice, direction, or criticism. Subsequently, he was to chair the WHO Advisory Committee on Post-Eradication Policy from 1981 to 1986 and was the senior author of the WHO archival history, *Smallpox and Its Eradication.*

By the time of the first meeting of the global commission, international commissions had certified eradication in forty-four countries, and the remaining ones were on schedule to be completed in 1979. Except for the first reviews in South America, the quality of programs and reports had been consistently satisfactory.

One disquieting question that remained was the status of smallpox in China, which had become a member of WHO in 1972. Little information was available on the occurrence of the disease, but health officials insisted that eradication had been achieved in 1960 by means of a mass-vaccination campaign. Chinese government officials believed that the official statement that the country had been smallpox-free for nearly two decades was enough information to provide. However, the commission took note of the vast size of China and the fact that it then constituted one-quarter of the world's population. Members decided that they could not conclude that global eradication had been achieved without concrete data from China.

Beginning in 1977, Arita and Director-General Mahler increasingly pressed Chinese officials for more information, using both formal and informal contacts with the government, but to no avail. Finally, WHO asked Sir Gustav Nossal, an internationally respected Australian scientist, to visit China to explain personally to senior ministry officials the importance of both a visit and a review of the program by representatives from the global commission. He was persuasive. The government agreed. In July, Fenner, chairman of the global commission, and Dr. Joel Bremen, WHO staff member, went to China (see figure 68). They were presented with remarkably complete documentation covering the history of the Chinese smallpox program, including epidemiological data for each province. The last endemic cases had occurred in 1961, but small numbers of additional cases due to variolation were detected over the next four years in three different counties of the remote Xizang Autonomous Region. To verify that this problem had been resolved, Chinese health officials conducted a large-scale scar survey in this region in the spring of 1979. There was no evidence of cases since those detected fourteen years before.

The global commission held its final meeting in December 1979 and reached agreement that the global eradication of smallpox had been

Figure 68. Drs. Frank Fenner and Joel Bremen in China. Bremen (*far left*) and **Fenner** (*second from left*) with Chinese officials in Kunming, China, 1979, during review of its smallpox program history. *Photograph courtesy of J. Bremen.*

achieved. A report for the director-general was prepared and a certificate signed by the members of the commission (see plate 7).

THE WORLD HEALTH ASSEMBLY—1980

On May 8, 1980, a special plenary meeting took place at the World Health Assembly in Geneva. Health ministers from nations around the world attended. Fenner presented the final report of the global commission and its recommendations (see figure 69). A formal resolution and recommendations were adopted by acclamation and signed by representatives of each of the governments. The first two recommendations were of special significance:

The Thirty-third World Health Assembly, on this the eighth day of May 1980:

1. Declares solemnly that the world and all its peoples have won freedom from smallpox, which was a most devastating disease sweeping in epidemic form through many countries since earliest time, leaving death, blindness and disfigurement in its wake and which only a decade ago was rampant in Africa, Asia and South America;

2. Expresses its deep gratitude to all nations and individuals who contributed to the success of this noble and historic endeavor;

3. Calls this unprecedented achievement in the history of public health to the attention of all nations, which by their collective action have freed mankind of this ancient scourge and, in so doing, have demonstrated how nations working together in a common cause may further human progress (see plate 8).

Recommendation: Smallpox vaccination should be discontinued in every country except for investigators at special risk.

Recommendation: International certificates of vaccination against smallpox should no longer be required of any travelers.

It was a moving ceremony and a memorable moment, but I felt an odd sense of loss. Never again could I expect to share with so many colleagues from so many countries the day-to-day excitement, stress, anxiety, elation, and satisfaction in achievement that had been our common experience for more than a decade.

POST-ERADICATION

The global commission made a number of additional recommendations for action, which were endorsed by the World Health Assembly. The principal recommendations and the ensuing activities are summarized below:

Figure 69. World Health Assembly, 1980—Dr. Frank Fenner Speaking. As chairman of the WHO Global Commission for the Certification of Smallpox Eradication, Fenner summarized its report for the assembly. The assembly adopted a formal resolution proclaiming that the world had won freedom from smallpox. *Photographs courtesy WHO/T. Germain.*

The book and the archives

"Recommendation: WHO should ensure that appropriate publications are produced describing smallpox and its eradication and the principles and methods that are applicable to other programs. All relevant scientific, operational and administrative data should be catalogued and retained for archival purposes."

An archival history seemed to be the best answer to meet this charge. Extensive documentation, data, and pictorial material had been amassed during the program and these were duly archived. A book was an added option—a massive undertaking. This required the knowledge of those of us who had participated in the program. It would be difficult for one not intimately involved to glean the essential elements of the complex, multi-faceted campaign and to fit the pieces together to make a coherent whole. We also recognized that in all likelihood, interest in smallpox would soon vanish into the mists of the past—remembered perhaps as only one of an array of diseases with rash, such as measles and chicken pox. Spurred on by Fenner, the authors embarked on a seven-year odyssey—the authors being Frank Fenner, Isao Arita, Zdeno Jezek, Ivan Ladnyi, and myself. We were greatly aided by an outstanding editor from WHO, Dominic Loveday, and the special assistance of John Wickett, one of our former administrative officers (see figure 70). Of the authors only Fenner and I had English as our native language and thus we became the principal writers. Arita, Jezek, and Ladnyi contributed drafts of chapters and comments. Copies of each of the operational chapters were circulated in draft to a number of participants in each of the country programs.

We recognized that if the book were to be instructive, it would have to present both successes and failures in the program and reflect both good and poor performances of participants. In brief it could not be an "institutional history." The director-general concurred. There was only one instance in which the authors deleted material at the insistence of a gov-

Figure 70. *Smallpox and Its Eradication*—Editorial Board and Principal Staff. March 26, 1986—(*left to right*) A. D. Loveday (WHO), K. Wynn (WHO), I. Arita (Japan), N. Henderson (USA), S. M. Deck (WHO), F. Fenner (Australia), I. D. Ladnyi (USSR), Z. Jezek (WHO), D. A. Henderson (USA), J. F. Wickett (WHO). *Photograph courtesy WHO/T. Farkas.*

ernment. There was a short section explaining the origin of the "Realm of the Final Inch," which was written by Aleksandr Isayevich Solzhenitsyn. At the time the book was being completed, Solzhenitsyn's writings were banned by the Russian government, and Ladnyi was passionately insistent that this material not be included. We reluctantly agreed.

The fifteen-hundred-page book, *Smallpox and Its Eradication*, included numerous photographs in color. It was published by WHO in 1988 and distributed widely to national governments, libraries, and academic centers. The book received a number of favorable reviews but not a great deal of attention—until 2001. The terrorist attack on the World Trade Towers in New York aroused new concern about the possible use of smallpox virus as a biological weapon. Remaining copies were quickly sold out. The book can now be accessed, however, on the World Health Organization Web site.

Monkeypox—a potential threat?

"Recommendation: In collaboration with country health services, WHO should organize and assist a special surveillance program on human monkeypox, its epidemiology, and its ecology in areas where it is known to have occurred."

An important unknown from the beginning of the program was whether smallpox might have an unsuspected animal reservoir. If so, it seemed likely that it would be in primates. Two previous global eradication programs—against malaria and against yellow fever—had revealed unexpected reservoirs in primates for each of these organisms. Could this be the case with the smallpox virus? There was no evidence for a natural reservoir, but it was crucial to confirm this.

Monkeypox was first discovered in 1959 in a Danish laboratory. A cynomolgus monkey had pustular lesions very similar to smallpox; from these a virus had been isolated called "monkeypox virus." It was closely related to smallpox virus, but it was a distinctly different entity. Had other laboratories seen such cases and had there been laboratory-infected human cases? Letters were sent to a large number of research scientists seeking information. A few laboratories reported occasional cases in monkeys, some of which came from Asia and some from Africa. No cases were known to have occurred among monkeys in the wild. We needed to know more, and so in 1969, Arita convened a special WHO consultant group with Frank Fenner as chair to plan pertinent research.

A significant goal was to determine whether monkeypox was present in the wild in Asia or Africa or both. This proved difficult to assess. Studies to measure serum antibodies in captive monkeys were of no help. It was found that monkeys, in shipment from different countries, were often kept in common holding areas and readily spread infection to each other. Serological surveys of animals in the wild likewise provided little help. Many different types of poxviruses exist in nature, and serological studies could not differentiate among these.

A major surprise had come in 1970 when, in Zaire, Ladnyi with Dr. Pierre Ziegler investigated and confirmed the first human monkeypox case. The subject was a nine-month-old boy who developed a fever and a rash that was indistinguishable from smallpox. He lived in a rural tropical rain forest and was the only case found in the area. The virus from this case was isolated by Dr. Svetlana Marennikova (see figure 71) in the WHO Collaborating Laboratory in Moscow. At the same time, Drs. James Nakano and Joseph Esposito (see figure 71) at the CDC laboratory in Atlanta were puzzling over smallpox-like viruses from two suspect smallpox cases in what was thought to be smallpox-free Liberia. Marennikova telephoned me to tell me about her findings and the methods she had used for diagnosis of

Figure 71. Drs. Svetlana Marennikova, James Nakano, and Joseph Esposito. **Marennikova** directed the WHO Collaborating Center on Smallpox in Moscow, and **Nakano** directed the one at the CDC. **Esposito** was a senior research member on the CDC team. They processed many thousands of specimens from countries throughout the world and undertook numerous research studies. *Photographs courtesy of S. Marennikova/CDC.*

the virus. I immediately called the CDC, which checked their viruses in the same manner. They discovered that the Liberian cases were also human monkeypox cases.

Subsequently, many additional human cases were reported from tropical rain forest areas of Africa, almost all being from Zaire. No victims had been vaccinated previously against smallpox. The death rate of about 10 percent was similar to that seen in African cases of smallpox. The cases were scattered widely throughout the rain forest areas. We surmised that monkeypox cases may have been widely prevalent for many years but could be identified as such only after smallpox had been eliminated. However, the monkeypox virus did not spread readily even to household contacts. Field studies pointed to small rodents regularly eaten by native peoples as the primary source of infection. In brief, monkeys, like man, were only occasional victims—the rodents were the endemic source.

We wondered how rapidly monkeypox virus might spread when vaccination ceased and vaccination immunity waned. Until this was clarified, we remained apprehensive that this discovery might represent another instance in which, as with yellow fever, there was an unsuspected forest reservoir. Following the declaration of eradication, an intensive WHO-sponsored five-year program of study was initiated under the direction of Zdeno Jezek and former WHO Zaire smallpox adviser Mark Szczeniowski. From 1982 to 1985, an average of seventy cases per year was found in a population study area of 125,000 people. Fortunately the outbreaks were small and died out quickly without vaccination or special efforts to isolate patients. Further studies are continuing.

Whether to reinstitute vaccination throughout the tropical rain forest was often discussed, since smallpox vaccine protects against the acquisition of monkeypox. However, vaccination throughout the vast rain forest area would be extremely difficult, security being a major problem. And with AIDS being widespread, the likelihood of serious adverse reactions following vaccination would be enhanced. For the present, only surveillance is being pursued.

Reserve stocks of vaccine

"Recommendation: Sufficient freeze-dried smallpox vaccine to vaccinate 200 million people should be maintained by WHO in refrigerated depots and periodically tested for potency."

At the time eradication was declared, it seemed unlikely that the world would ever again need large quantities of smallpox vaccine. There was no known smallpox reservoir in nature, and the virus remained in only two laboratories so far as we knew. In 1980 no thought was given to the potential use of smallpox virus as a biological weapon—except, as we learned many years later, in the Soviet Union. The figure of 200 million doses for storage was proposed simply because there was that amount in storage in WHO's rented -20°C (-4F°) cold storage locker in Geneva. The costs for continuing storage and periodic retesting were only about $25,000 per year.

The vaccines had been donated by several different countries and their potency, purity, and stability had passed quality control tests in the producers' laboratories as well as at the WHO vaccine reference laboratory in the Netherlands. Stability tests of smallpox vaccine had shown that most vaccines retained potency for at least ten years.

During the mid-1980s, WHO experienced financial problems necessitating cutbacks in many different programs. A WHO ad hoc smallpox advisory committee was consulted about reducing the stocks. The committee advised that the vaccine in storage be reduced to five million doses and that the balance be destroyed or returned to the donor countries if they requested it. This was done.

In 2003 WHO convened the advisory committee again. By that time it had become known that the Soviet Union had been actively engaged in a biological weapons program and had given priority to weaponizing smallpox virus. Whether other countries might have similar intentions was unknown. Smallpox was a special concern because of its capacity to spread—especially in a now highly susceptible, largely unvaccinated global population. It was agreed by the committee that a substantial stockpile of vaccine for international emergency use was essential. It reversed the 1985 recommendation of the advisory committee and urged an increase of the emergency stockpile to at least 200 million doses. The United States committed 20 million doses, but substantial additional donations from other countries have yet to be received.

Laboratories retaining stocks of smallpox virus

"Recommendation: No more than four WHO collaborating centers should be approved as suitable to hold and handle stocks of variola virus.... Each such center should

report relevant information on its safety measures annually to WHO and be inspected periodically by WHO."

With cessation of smallpox vaccination and with increasingly susceptible populations, the implications of an accidental escape of smallpox virus from a laboratory were increasingly troubling. The recommendation that the virus be restricted to only four laboratories and that the laboratories and their safety measures be regularly inspected by WHO were important and precedent-setting.

In 1976 WHO had undertaken a survey of 823 laboratories then on the WHO World List of Virus Laboratories to ascertain which ones had processed smallpox specimens and might be retaining them. The questionnaire also asked if those retaining the virus would agree to destroying their stocks or to transferring them to a WHO collaborating laboratory Virtually all countries responded—seventy-five laboratories replied that they were holding smallpox virus samples, and most indicated a willingness to destroy or transfer the material.

A WHO expert group on laboratory safety issued in 1976 the first official statement describing the kinds of physical containment provisions necessary for handling dangerous pathogens, more specifically smallpox virus. These called for air and biowaste sterilization, glove boxes, two-way autoclaves, and biological safety cabinets.

Soon thereafter, concerns about laboratory safety increased dramatically as the result of an accident at the University of Birmingham, England. On August 11, 1978, Janet Parker, a forty-year-old medical photographer at the university, became ill with fever and developed a rash four days later. She had been vaccinated twelve years before. She had not traveled prior to becoming sick and she had no known contact with anyone with an illness with rash. The world's last case of smallpox had occurred in Somalia fully ten months before and little thought was given to smallpox as a possible diagnosis. However, a specimen obtained on August 27 revealed the smallpox virus.

The university housed a smallpox research laboratory under the direction of virologist Dr. Henry Bedson, who was collaborating with WHO on studies of various smallpox virus strains. Directly above his laboratory was a telephone booth used by Ms. Parker. An air duct connected the laboratory with the booth. Presumably, it was via this duct that she became infected with an aerosol of smallpox virus.

Ms. Parker was promptly isolated, but she infected one other person, her mother, before her death on September 11. Her mother developed only a few lesions and recovered. Professor Bedson, totally distraught by the events, committed suicide, dying on September 7.

Health administrators in many countries called for the end of laboratory studies of variola virus and the destruction of all stocks of the virus. National government officials discouraged their own research laboratories from retaining the virus. The number of laboratories with the virus dropped from seventy-five to four by the time of the May 1980 declaration of eradication. Subsequently, laboratories in the United Kingdom and South Africa parted with their smallpox specimens. The only remaining stocks of the virus were at the Centers for Disease Control in Atlanta and at the Research Institute for Viral Preparations in Moscow.

TO DESTROY OR TO RETAIN THE REMAINING STOCKS OF SMALLPOX VIRUS

The WHO Global Commission in its first meeting in 1979 recommended that "[a]n expert group to report to the Global Commission should be convened by WHO to investigate whether retention of stocks of variola virus is justified after global eradication has been completed and, if it is justified, to identify the need for and nature of any research to be conducted." Meanwhile, of course, the vaccinic virus would continue to be retained.

The question of possible destruction of the last stocks in the two reference centers was taken up by the WHO Advisory Committee on Post-Smallpox Eradication Policy on which I served. The committee met annually from 1981 to 1985. Until the 1990s, however, neither I nor other advisory committee members felt that there was urgency in addressing the problem. As far as we knew, smallpox was present only in the two laboratories, and neither was doing research using variola virus. Both were directed by competent, knowledgeable scientists.

At the World Health Assembly, representatives from the recently endemic countries pressed for the final destruction of the virus. They pointed out that they were the ones that had most recently endured the impact of epidemic smallpox. They wanted all possible steps to be taken to ensure that there would not be a recurrence of smallpox. They argued that the decision to undertake the global eradication of smallpox had been taken by the World Health Assembly as a whole; that the most substantial contributions to that effort had been made by the endemic countries themselves; and that the World Health Assembly should have the opportunity to debate and decide this issue. An underlying suspicion permeated the questions: "Why is it that only the two major powers are keeping the virus and no one else? Is it possible that they have malicious intentions for eventual use of it?"

The initial steps toward smallpox virus destruction

The first concrete initiative of the WHO advisory committee toward the possible destruction of the virus was taken in 1985. The committee sought the views about variola virus destruction from sixty virologists in twenty-one countries; only five expressed the view that the virus should be retained indefinitely. Accordingly, in 1986, the committee recommended that work begin to preserve viral genetic information should it be needed at some time in the future. The two reference centers worked together to create libraries of noninfectious cloned fragments of the principal strains so that studies could be conducted without risking escape of the virus.

Interest in virus destruction heightened significantly in 1990 when Secretary of Health and Human Services Louis Sullivan, in an address to the World Health Assembly, announced the intention of the United States to destroy all smallpox virus stocks, following the successful sequencing of the virus genome of several different strains. He anticipated this would require about three years. He invited the Soviet Union to endorse this effort, and it agreed. The advisory committee concurred in the proposal and a tentative December 31, 1993, target date was set.

In light of subsequent events, it is incumbent to point out that the intentions, the deliberate steps taken, and the rationale were widely publicized. The advisory committee invited open international debate. The pros and cons of destruction were argued at an August 1993 International Congress of Virology in Glasgow, Scotland, and summary papers were subsequently published. WHO then solicited the views of five scientific groups. In 1993, written endorsements for destroying the virus were received from each: the executive board of the International Union of Microbiological Societies, the Council of the American Microbiological Society, the

Council of the American Type Culture Collection, the Russian Academy of Medical Sciences, and the board of scientific counselors of the CDC's National Center for Infectious Diseases.

Meanwhile, in March 1993, with the advent of the new Clinton administration, I had resigned my White House post as associate director of the Office of Science and Technology Policy to become deputy assistant secretary for science and health in the Department of Health and Human Services (DHHS). This placed me at the center of discussions deciding US policies and programs regarding smallpox—and in implementing the apparently straightforward steps to be taken in a now well-discussed program to destroy known stocks of smallpox virus. At that time I had no inkling of the time and effort that would subsequently be consumed in a continuing series of contentious and sometimes irrational discussions that are now planned to continue through at least 2011.

Sequencing of the genome was completed in January 1994. At a meeting that month, Dr. Bernard Moss of the US National Institutes of Health (NIH), a leading scientist in pox virology, stated, "Now we are satisfied that the genetic blueprint of variola virus has been properly achieved for posterity." In September, the advisory committee reviewed again its recommendation on destruction. Eight of the ten members, including myself, voted for destruction of the virus on June 30, 1995; two members argued for a further two-year delay. A final report for the director-general was completed and scheduled for presentation to the executive board of the World Health Assembly in January 1995. Endorsement by the assembly was expected in May.

Objections to virus destruction

Between 1986 and 1994, post-eradication plans and programs proceeded uneventfully. This changed in November 1994. A senior official of the British Chemical and Biological Defense Establishment at Porton Down asked to meet with me. He was at that time visiting Washington. I had not met him before, but he knew that I had been a member of the WHO advisory committee dealing with smallpox eradication policies. He came directly to the point in inquiring how it was that the committee could advise that the smallpox virus be destroyed. I asked what difficulty he foresaw. He said this could be disastrous for the world. Suppose, he said, that someone were to exhume a frozen smallpox corpse from the Arctic tundra, that the virus then spread, and there was no way to stop it? Periodically, there had been speculation that smallpox virus might be preserved in a body kept at a very cold temperature. However, even if a virus did infect those exhuming the corpse, that was not a major problem. An outbreak could readily be controlled with smallpox vaccine. After all, there was no plan to destroy the smallpox vaccine.

Despite the responsibilities of this British official, he seemed unaware of the difference between smallpox virus and the cowpox-derived vaccine virus. This puzzled me. In England especially, the story is well known of Edward Jenner's historic discovery of smallpox vaccine—a demonstration that infection caused by pustular material from a cowpox lesion protected against smallpox. I carefully explained to the director the differences between the two viruses and the fact that there was no thought of destroying the vaccine virus. As he departed, he said he had no other reservations about the recommendations of the WHO committee. However, this meeting was only the opening round of deliberations that became ever more strange.

Prior to the January WHO executive board meetings, discussions took place to determine what policies the US delegates should and should not support. Since 1990 the United States had not only consistently supported but had actively advocated destruction of the smallpox virus. The only action proposed for the 1995 board was to endorse the report of the advisory committee and to reaffirm what the United States had proposed and the Soviet Union had agreed upon four years before. They would also have to approve the proposed date of June 30, 1995, for destruction of the virus.

Within days after my meeting with the British director, I was called to a special committee meeting of the National Security Council at the White House. In addition to White House staff, there were representatives from the Departments of State, Defense, Commerce, and DHHS, as well as the Arms Control and Disarmament Agency. On the agenda was the question of the policy position to be taken on smallpox virus destruction by US delegates at the January WHO Executive Board. (The board, composed of thirty-two country representatives, develops policy recommendations for the full assembly.) One by one, each of the representatives, on behalf of his agency, agreed to support the established US position—except for the Department of Defense (DoD) representative, who registered a vehement "non-concurrence." He was asked to explain. Astonishingly, he repeated the "bodies in the tundra" threat. I said that this was an erroneous concern and explained the difference between vaccinia and variola viruses. Other objections were raised that derived from a fundamental lack of understanding of the biology and epidemiology of smallpox. These I dealt with one by one.

Eventually, the White House chairperson closed the meeting, indicating her hope that some sort of compromise could be reached. Meeting after meeting was to follow, continuing into January.

Why and how, after four years of open, public debate and discussion, an established government policy should be abruptly derailed without a plausible explanation was a mystery. All were in general agreement that an argument could be made to retain the intact smallpox virus for possible research purposes not presently foreseen. This had to be weighed against the threat of the possible escape of the virus. The WHO advisory committee and the five scientific organizations had come down on the side of destruction.

Nobel laureate David Baltimore, then president of Rockefeller University, summed up the arguments in a letter of January 7, 1994, in *Science* magazine:

> I find it hard to believe that we need to, or even will, continue research of a virus whose release from containment would be such a disaster while its present threat is nil.... While I agree that a deeper understanding of pathogenesis will help counter microbial infections, I doubt that we so desperately need to study smallpox that it would be worth the risk inherent in the experimentation. Much of the value of research can be gained from studying related viruses, especially vaccinia. Eradication of the virus as well as its disease will better serve the long-term interests of humanity as the proponents of destruction have argued.

Other arguments were advanced by those whose primary expertise seemed to be nuclear weapons. What would happen if we destroyed our stocks of virus and the Russians didn't destroy theirs? I asked if this implied that we would want to retaliate in kind to a biological weapons attack with smallpox. All were certain this would never be the case. A "constructive" suggestion was offered—since the two countries each had several hundred different strains of smallpox (a strain being the virus isolated from one patient), we might work out an agreement for the United States and Russia each to destroy all but perhaps fifty strains. Regarding a biological agent, this made absolutely no sense: one could grow all the virus one needed from the contents of one tiny vial. One could imagine the countries hurling vials of virus back and forth. Such was the tenor of discussions.

The January 1995 executive board is blocked from taking action

The US delegation to the January 1995 executive board eventually went with instructions to work with other delegations to avoid having even a discussion by the board regarding destruction of the virus. It was widely known that a majority of members of the WHO executive board were in favor of destruction, as were those in the World Health Assembly. The US officials believed that a formal vote in favor of destruction would put the United States in a difficult position.

At the executive board, behind-the-scenes lobbying by US and British officials arranged to have the advisory committee report dismissed without discussion on the grounds that a consensus did not yet exist, that it was an issue that was too important to settle by majority vote. This action in response to a report of a WHO scientific advisory committee was unprecedented.

From a practical standpoint there were serious difficulties in working with smallpox virus in the laboratory—one of the reasons that there had been little experimental work. The virus could be grown and studied only in tissue cell culture because it did not infect animals. Moreover, studies had to be conducted in one of two high-security laboratories, either at the CDC or in a Russian facility. One other US high-security laboratory was a DoD facility in Frederick, Maryland, but military officials feared possible protests if studies with smallpox were to be conducted there. Thus, use of this laboratory was not further considered.

Historically, most of the studies exploring the possible pathogenesis of smallpox virus had been conducted using other members of the ortho-poxvirus family—monkeypox, mousepox, rabbit pox, vaccinia, cowpox, and others—all of which grow in mammals. The possible escape of one of these viruses posed no threat to the community. For determining how new smallpox vaccines might protect humans, vaccinated monkeys were chal-lenged with monkeypox virus. This experiment was especially useful because monkeypox is closely related to smallpox and causes a disease in humans that is clinically similar to smallpox. For evaluating antiviral chemical agents for use in treatment, tissue cell cultures were used for initial screening. But information derived from cell cultures was only a first step, it had to be validated by using other orthopoxviruses in monkeys or other animals.

As time passed, I learned more about the extensive and possibly still extant Soviet biological weapons program and recognized the possibility that expertise and perhaps strains of virus might have been transferred to other countries. It also had to be recognized that the virus itself eventually might be synthesized. Thus, we could never be absolutely certain that no laboratory, anywhere, possessed smallpox virus. Consequently, steps are needed to decrease the likelihood of misuse of the smallpox virus. One approach would be for the World Health Assembly and United Nations General Assembly to take the formal position that after the designated date of smallpox virus destruction, any scientist or any country found to possess smallpox virus would be deemed guilty of a crime against humanity.

Attempts to reconcile an impasse

I hoped that through a reasoned, objective examination of the basic facts and issues, we might find common ground between the apparently irreconcilable positions of the DoD and DHHS. I proposed that two committees be created. One would be composed of external experts, half to be appointed by the DoD and half by DHHS, to examine the scientific and policy issues pertaining to virus destruction. Members were drawn from the Armed Forces epidemiological board and the CDC's board of scientific counselors. A second committee would include scientists from the two departments who would delineate specific research projects to undertake. They planned to commence their studies immediately. By the end of the year, they would be better able to estimate what more was required. The second committee was chaired jointly by myself and a representative from DoD.

The external policy committee identified three research areas that would require the use of live variola virus, but none were deemed to be of high priority. They believed that each could be addressed in a comparatively short period of time, provided the effort was given sufficient priority and resources. The committee conceded that a short postponement of virus destruction might be needed to fully clarify the requirements.

However, they proposed that a fixed early date for virus destruction be established.

The second committee, with knowledgeable virologists from both departments, began research studies of the nature they had identified as being needed. Monthly progress reports were submitted, and, in December, the investigators announced that the studies had been completed as agreed upon and that there was no further need for retaining the virus.

The result of all of this—no change. In retrospect, it was naive on my part to believe that reason and science might prevail.

The executive board and the World Health Assembly—1996 and onward

With the January 1996 executive board on the horizon, the US position was revisited. The DoD remained intransigent. Thus, the National Security Council staff asked that President Clinton adjudicate the contrary views. He decided that the United States should support the recommendation of the WHO advisory committee that the virus be destroyed in June 1996. However, he authorized the US representative, if necessary, to join in a consensus with other members of the board to postpone destruction for "rational reasons."

After extended discussions, the 1996 WHO executive board and World Health Assembly adopted resolutions to destroy the virus, but, as a compromise, to postpone the target date for destruction to 1999.

In preparation for the 1999 board and assembly meetings, the US agencies asked the Institute of Medicine (IOM) to convene a committee to review "the potential scientific and medical information that would be lost were live variola virus no longer available." The charge was a narrow one limited to yet again exploring possible research studies that might be conducted using the live virus. The committee was specifically told not to address the question of whether virus stocks should be retained. It was also not to assess the risks of retaining the stocks of virus or the possible costs or priorities that should be given to the research, or the possible alternative use of other orthopoxviruses.

The 1999 published IOM report contains a thoughtful list of possible studies that might relate to the development of new vaccines, antiviral drugs, better diagnostic methods, and an understanding of immune mechanisms. The concluding sentence of the report states, "The risks of retaining the stocks of live variola virus might well outweigh the benefits. If the stocks were retained, however, they could offer the possibility of scientific advances that could not otherwise be achieved." These were essentially the same conclusions reached by the WHO committee in 1986, thirteen years before.

Also preparatory to the 1999 WHO discussions, the director-general asked all member countries to express their official views about smallpox virus destruction. Replies were received from 79 of the 190 countries. One (Russia) voted for retention of the virus; four (United States, France, Italy, Great Britain) indicated that they had not yet decided; seventy-four countries said it should be destroyed.

When the assembly reconsidered the question in 1999, the United States requested an indefinite postponement for virus destruction, but delegates insisted that if the virus was to be kept, that it be retained for no more than three years. The assembly decided that a special WHO advisory committee on research should be constituted to meet annually. It was charged with the responsibility of approving all research protocols before they were undertaken in either the Atlanta or Moscow laboratories. The resolution also called for inspectors to evaluate each laboratory at least every two years to ensure that they were in full compliance with all security procedures.

Intrusive requirements such as these were unprecedented, but members of the board and the assembly remained deeply concerned about an accidental release of the virus. They were also wary about the uses to which the virus might be put. Suspicion had heightened when, unannounced, the Russian government moved its stock of variola virus from Moscow to the State Center for Research on Virology and Biotechnology (called VECTOR) in Koltsovo, Novosibirsk Region. VECTOR had been the Soviet Union's principal bioweapons research laboratory for viral agents. This included smallpox.

Virus destruction was again debated at the board and assembly in 2002. Prior to the meetings, the United States launched a vigorous effort to try to persuade countries not to set time limits but to permit the virus to be retained until specified goals had been met. They suggested the following goals: (1) licensure of two antiviral drugs for the prevention and treatment of smallpox infections; (2) licensure of a new vaccine that protected all individuals against all strains of variola and that were totally safe; (3) special environmental detectors for identifying and containing outbreaks of smallpox; (4) new technologies to deal with genetically modified strains of smallpox. It was widely recognized by scientists that such goals could not be achieved for decades, perhaps ever. The United States also sought agreement to abolish the requirement that a WHO research advisory committee approve all research protocols. The committee had so far denied permission for about half of the proposed protocols on the grounds that they carried too much risk or were not time-limited. The US initiative had effectively no support.

The 2002 assembly authorized further retention of the virus provided that the research was "conducted in an open and transparent manner only with the agreement and under the control of WHO." Further, it stipulated that the research be "outcome-oriented, time-limited and periodically reviewed." An annual report to the assembly was agreed upon.

Beginning in 2006, increasing demands began to be made by developing country representatives for setting a final, fixed date for virus destruction. They argued that ample time had already been allowed for research and that it was time to stop further research. They proposed that the year 2010 be established. The United States again led the opposition. Meeting after meeting was convened, but no consensus could be reached. There seemed to be no middle ground between having a fixed date for destruction and no target date at all. What is now planned is a major report to be prepared for discussion and review at the 2010 assembly with final resolution at the subsequent assembly in 2011.

Meanwhile, research programs using variola virus continue both at the CDC and at VECTOR. Two new vaccines are available: a Japanese-developed attenuated *vaccinia* strain called Lc16m8 and a German-US product called MVA (Modified Vaccinia Ankara), also an attenuated *vaccinia* strain. Neither vaccine can be said to meet the US stipulation of being "totally safe," nor can it be known for certain that either vaccine would actually protect humans against smallpox. There is no way to test this without infecting humans with variola virus! Work continues in developing an antiviral drug for treatment of smallpox cases, but no product has yet been shown to be effective in monkeypox-infected animals if given after they begin to develop fever and rash.

By 2011, more than fifteen years will have elapsed since work with the live smallpox virus was restarted in the United States after a hiatus of more than a decade. As time has passed, both the budget for research and the number of engaged laboratory personnel have steadily increased.

There appears to be no end in sight. The story of smallpox has a life of its own. I had concluded on more than one occasion that the time had come to relegate it to history—only to have new hurdles arise. For the first time ever, we have succeeded in eradicating a human disease. But eradicating the virus itself is another problem. Scientific and political issues are intertwined and seem unlikely to be resolved in the foreseeable future. It has been more than forty years since I first confronted the challenges posed by smallpox. I never dreamed that I would still be preoccupied with it.

The saga continues, but lest we forget—more than thirty years have elapsed since there has been a case or a death due to smallpox; complications caused by vaccinations have ceased; travelers move easily across borders no longer needing certificates to verify successful vaccination within the preceding three years.

But smallpox eradication was the genesis of something far greater. It gave life to what is called the Expanded Program on Immunization (EPI)—the fruition of our efforts to expand smallpox vaccination activities to include, at first, vaccines against poliomyelitis, measles, diphtheria, whooping cough, and tetanus. Such programs are now to be found throughout the world. Other vaccines have since been developed and are being applied; the research agenda promises many more. Prevention through immunization is a field and a future with vast, unimaginable horizons.

Chapter 10

SMALLPOX AS A BIOLOGICAL WEAPON

By 1990 I believed the shadow of smallpox had finally lifted and the disease was consigned to history. I mentally closed the smallpox chapter of my life. Our "Big Red Book," *Smallpox and Its Eradication*, had been published, and for most people the disease was a fading memory. The only remaining question, I thought, was to decide when the last stocks of smallpox virus should be destroyed. I was wrong. Soon, smallpox—the threat of it—was back on the international agenda, as ominous as ever.

The basis for our concern was a newly revealed secret biological weapons program in the Soviet Union. Smallpox virus was at the top of its priority list. The first certain intimation of the program came in 1989 from a Soviet defector, Dr. Vladimir Pasechnik, who reported conducting experiments on biological agents in his Leningrad laboratory. He was aware that other laboratories were similarly involved. This was clearly in direct defiance of the 1972 Biological Weapons Convention, which called for all offensive biological weapons to be destroyed and all research on their usage for offensive purposes to be stopped. All the major countries were signatory to the agreement, including the Soviet Union.

In 1991, after fourteen years as dean at Johns Hopkins, I had returned to government—serving in the White House as associate director for life sciences in the Office of Science and Technology Policy. During my tenure in the White House, concern about biological weapons was not high on the agenda. Some knew of the Pasechnik defection, but many remained skeptical about the likelihood of a major Soviet program. A prevalent belief was that no country would dare to use biological weapons in any case because of fear of possible nuclear retaliation. Moreover, most believed that biological weapons would be difficult to produce and disperse.

Additional information received during the mid-1990s suggested that the Soviet network of biological weapons research and production laboratories was extensive. This raised serious questions: What should the United States do about it, and how should the world community respond? What were the organisms of greatest interest and how might they be dispersed? By 1997 I had become increasingly engaged in addressing these questions. Smallpox was of special concern. Not only did it have a 30 percent death rate but it now could spread more readily than ever in a largely unprotected world population. Routine vaccination in the United States had stopped in 1972 and in the world in 1980. The immunity among those previously vaccinated was fading or even absent. The threat was especially ominous because now there were no smallpox vaccine production laboratories and only small amounts of vaccine in frozen storage.

Following the terrorist attacks on the World Trade Towers and the Pentagon on September 11, 2001, and shortly after the anthrax attacks, the US government began to strengthen many components of national security, including an intensified effort to cope with the biological weapons threat. Shortly after the September 11 disaster, I was asked by Secretary Tommy Thompson of the Department of Health and Human Services (DHHS) to organize and direct a newly created national Office of Public Health Emergency Preparedness (later renamed the Office of Preparedness and Response). It was to be headed by a newly created post of assistant secretary.

National perceptions, attitudes, and policies were undergoing a radical transformation. Once again, I was coping with a broad agenda, but a high priority was to devise strategies to prevent the spread of smallpox should an attack occur.

BIOLOGICAL WARFARE—THE EMERGING THREAT

Until the mid-1990s the United States had paid little attention to the potential use of biological weapons, either by another nation or by terrorists. It was assumed that all countries, including the Soviet Union, were complying with the biological weapons ban. However, the 1972 convention had made no provisions for inspections to ensure that countries were not violating its stipulations.

Biological warfare is not new. It has been used in various forms for centuries—even to spread smallpox. During World War II the Japanese developed an elaborate biological weapons program. They did not use the smallpox virus, but they employed a number of other agents such as plague and cholera to attack Chinese civilians. Until the 1972 convention, a number of countries, including the United States, conducted their own national programs, fearing that weapons similar to those used by Japan might be unleashed against them. The convention agreements had been

---

SMALLPOX AS A BIOLOGICAL WEAPON—EIGHTEENTH CENTURY

During the French and Indian Wars in North America, Lord Jeffery Amherst proposed that smallpox be used to defeat the Indians. In a July 7, 1768, letter to Colonel Henry Bouquet, who was stationed at Fort Pitt, Amherst wrote, "Could it not be Contrived to Send the Small Pox among these Disaffected Tribes of Indians? We must on this occasion, Use Every Stratagem in our power to Reduce them." On July 15, Bouquet responded, saying, "I will try to inoculate [them] with Some Blankets that may fall in their Hands, and take care not to get the disease myself." According to the diary of a trader stationed at Fort Pitt, the plan was carried out on June 24, 1763. Two Indian dignitaries visited Fort Pitt, hoping to persuade the British to abandon their posts. Unsuccessful, they left, requesting "a little provisions and liquor, to carry us home." The trader wrote, "Out of regard to them, we gave them two Blankets and an Handkerchief out of the Smallpox Hospital. I hope it has the desired effect." By the following spring, smallpox was said to be raging among all the Indians in the Ohio River Valley.

The US Revolutionary War developed soon thereafter. Historian Elizabeth Fenn documents a number of instances in which smallpox-infected patients were used by the British military to initiate outbreaks in the Continental Army. It was a useful weapon. British troops at the time were immune to smallpox because of variolation or natural infection. Variolation had seldom been practiced in North America. However, in 1777, General George Washington ordered the variolation of all his troops.

intended to remove, once and for all, the threat of biological agents and the fear that these weapons might spark catastrophic epidemics.

The Soviet Union's secret—a massive bioweapons program

The Soviet Union's sophisticated, clandestine biological weapons program developed rapidly after World War II. In 1947 the Ministry of Defense established a permanent "Center of Virology" in Zagorsk, a town northeast of Moscow. As we later learned, Soviet policy makers considered the smallpox virus to be the ideal weapon for inflicting large numbers of civilian casualties. Researchers tested numerous strains to determine which were the most virulent. Additionally, they tried various methods for dispersing smallpox virus in an invisible aerosol cloud. For years there were rumors of field tests with smallpox virus being performed on Vozrozhdeniye Island in the Aral Sea. This was confirmed in 2003 in recovered Soviet documents. A 1971 outbreak, believed to have begun by aerosol release, first affected a marine scientist on a boat that was passing the island—smallpox soon spread to a nearby town called Aralsk. In 2005, the Russian deputy minister of health publicly admitted that aerosolized smallpox virus experiments had been conducted during this period on the Aral Sea island.

The Zagorsk laboratory, the major Soviet manufacturer of smallpox virus, was capable of producing the virus in large quantities. Several tons were kept in cold storage and replaced annually. The intent was for the virus to be loaded into bomblets to be placed within intercontinental ballistic missiles (ICBMs). At a critical level of descent, the ICBMs would open a parachute to slow their speed, releasing the bomblets and dispersing the virus as an aerosol over the target area.

After the signing of the convention, the Soviet program was greatly intensified. Secrecy was more important than ever. A major part of the research component of the program was hidden within a state-owned pharmaceutical company called Biopreparat. After the World Health Assembly declaration in 1980 that smallpox had been eradicated, the Soviet Ministry of Defense approved a five-year plan (1981–1985) to undertake studies at Zagorsk of an improved, more virulent strain of smallpox virus.

At the time of the Pasechnik defection in 1989, the project had grown to include an estimated 60,000 people working in some fifty different laboratories. It is difficult to imagine that an enterprise this large could be effectively hidden. However, secrecy was extraordinarily tight and the capabilities of US and allied intelligence services in the biological weapons field were limited. Many of the initial revelations were so startling that their credibility was doubted. In the United States the information was classified at the highest levels and shared with only a small number of officials. Despite my high-level security clearance, I learned nothing definitive about this until after I left the White House.

In 1990, US president George H. W. Bush and British prime minister Margaret Thatcher discussed the Pasechnik allegations privately with the Soviet leader, Mikhail Gorbachev. He denied them emphatically. As a demonstration of good faith, he proposed that US and British scientists visit supposed biological weapons laboratories in the Soviet Union and Soviet scientists make reciprocal visits to institutions in the United States. This was agreed upon. Four laboratories each in the United States and Russia were designated.

In January 1991 a British-American team visited the State Center for Research on Virology and Biotechnology, known as "VECTOR," located in Koltsovo, in western Siberia near Novosibirsk. It was the premier laboratory of the secret Biopreparat operation, consisting of about fifty buildings and 4,000 workers. Officially, the facility was making pesticides for agricultural use. The chief scientist and first deputy director of Biopreparat, Dr. Kanatjan Alibekov (he later changed his name to Ken Alibek) was personally responsible for ensuring that the visiting team would find nothing that would indicate the laboratory was a bioweapons research site.

There was much to hide. Under Alibekov's direction, the laboratory had been engaged in a five-year development plan (1986–1990), the goal of which was to produce smallpox virus in quantities of fifty to one hundred tons a year. The laboratory was also endeavoring to make an "improved" smallpox virus by combining smallpox with a hemorrhagic disease virus to make it more lethal and an encephalitis virus to make it more transmissible from person to person.

VECTOR took extraordinary measures to prevent its workers from becoming infected: if infection were to occur and be reported, it would provide undeniable evidence in an otherwise smallpox-free world that forbidden research was in progress. The work was performed in a dedicated building. Workers were isolated for three to four months at a time; they ate and slept on-site. Before being allowed to visit their homes, they were held in quarantine for two weeks to make certain they were not infected.

During the inspection team's visit, British virologist David Kelly asked a technician what virus he was working on. He was told that it was smallpox. The question was repeated and retranslated with the same result. Kelly asked further questions about smallpox virus research—and then, abruptly, the tour of the laboratory was aborted. After an extended wait, the laboratory's director appeared and explained that they were only working on cloned fragments of smallpox virus that had been provided by the Moscow WHO Smallpox Collaborating Laboratory. However, the inspection team did discover special equipment in the laboratory that could be used to test methods for dispersing the smallpox virus. The team left with suspicions but without definitive evidence of banned activities.

In December 1991 Alibekov led a Soviet team on inspection visits to American laboratories—sites selected by the Soviet military intelligence (GRU) that it believed were involved in making biological weapons. At each laboratory Alibekov was said to have been an extremely aggressive, skeptical, probing investigator. Despite his efforts the Soviet team found no evidence of offensive biological weapons research. As Alibekov explained to me, he had been told by his government that the United States was making biological weapons. He had rationalized his role in the Soviet bioweapons program as a patriotic responsibility—to help his country counter the American threat. Alibekov said he was devastated to discover that there was no such threat, and at that point he began to contemplate defecting to the United States.

The Soviet team returned to Moscow and turmoil. A few days earlier President Gorbachev had resigned. The Soviet Union had become the Russian Federation, and Boris Yeltsin was president. In October 1992, with the help of the US Central Intelligence Agency, Alibekov was spirited away to the United States. His extensive debriefing lasted for more than a year as the intelligence agencies repeatedly questioned his accounts. The scope and sophistication of the program he described were beyond belief. Confirmation from other sources was needed. For more than three years, Alibekov's presence and revelations were known to few people outside a national security inner circle.

I do not recall having known about Alibekov until at least 1995. Even then, I had been repeatedly assured that he could not be believed; that he was fabricating information to obtain respect and special favors. Finally, I had the opportunity to converse at length with him, to read his book, *Bio-hazard* (published in 1999), and to talk with other Russian scientists. Some assertions in the book may be questionable, but Alibekov's detailed descriptions of the programs portray the disturbing reality of a massive and sophisticated biological weapons capacity.

The Soviet bioweapons legacy— who else might have the smallpox virus?

The breakup of the Soviet Union brought new worries, particularly about smallpox. The Russian government slashed its funding for the biological weapons laboratories. In some, including VECTOR, the number of employees was cut by at least half. Salary payments were irregular, and scientists began leaving for other countries. Many of them were highly trained professionals, experienced in producing large quantities of virus and knowledgeable of experimental work that had endeavored to combine the smallpox virus with other viruses. It was possible that some of these scientists might have taken samples of the smallpox virus with them. Also troubling was that recruiting teams from some Middle Eastern countries had visited VECTOR and other laboratories, offering generous consulting fees—and finding scientists willing to accept them.

In 1994, four Russian government biological weapons experts met privately with several scientists at the US National Academy of Sciences. Material they showed us explained how they had prioritized agents to assess which would make the best biological weapons. On their list were some twenty candidates: each was ranked from 1 to 10, based on characteristics of lethality, ease of spread, stability, the availability of protective vaccines or therapeutic drugs, and other characteristics. Smallpox topped this list by a wide margin despite an inferior mark for one characteristic— the availability of a good protective vaccine that could be used to stop an epidemic. The Russians had not taken into account that there was little vaccine available worldwide, that there were no smallpox vaccine manufacturers, and that it would require at least five years to produce any substantial amount of vaccine. Smallpox should have scored much higher on their priority list. I chose not to point this out.

Two other events in 1995 shook a complacent United States. First was the release of sarin gas in the Tokyo subway by an apocalyptic cult called Aum Shinrikyo. This cult—large, but until then little known—had also tried to disseminate anthrax organisms by aerosol throughout Tokyo. The attacks failed primarily because the group had used a strain with vaccine-like characteristics instead of a typical virulent strain of anthrax. Until this event, policy makers had assumed that such a sophisticated attack could be undertaken only by a national government—that it was beyond the abilities and resources of any private group. The US government believed that the threat of nuclear retaliation to any nation mounting such an attack would be an important deterrent. However, retaliation against a private group dispersed throughout the population, such as Aum Shinrikyo, would be impossible.

The second event was the defection of Hussein Kamal. Kamal, the son-in-law of Iraq's Saddam Hussein, provided detailed documentation to the United Nations describing Iraq's biological weapons program during the late 1980s. The so-called chicken coop papers (their site of storage) showed that Iraq had produced, filled, and deployed bombs and rockets containing anthrax organisms. They had also tested drone aircraft to spray anthrax. Nothing was said about smallpox. Later evidence confirmed that the program had been abolished in the early 1990s. However, it was an example of another nation with the intent to wage biological warfare—despite having signed the 1972 international agreement not to do so. How many others might there be?

AN ILL-PREPARED UNITED STATES AWAKENS TO A THREAT

In 1995 President Bill Clinton issued a Presidential Decision Directive to all agencies requesting that they begin to develop counterterrorism programs. In the following year the US Congress passed the Domestic Preparedness Initiative Act. It called primarily for training and supporting first responders in one hundred and twenty cities. First responders were identified as being police, firefighters, and emergency medical response teams. The Departments of Defense and Justice would be responsible for directing and funding the program. No provision was made for developing community-wide programs involving hospitals, physicians, or public health personnel. In fact, DHHS itself was given no role. This was a serious omission, as the Centers for Disease Control and Prevention (formerly the Communicable Disease Center but still called the CDC), along with state and local health departments, had been principally responsible for containing disease outbreaks. Yet, until 1999, the CDC had no professional staff whatsoever who were concerned with the public health response to a biological weapons attack.

Having become increasingly concerned about the specter of bioterrorism, I went to the army's Aberdeen, Maryland, facilities where first responders were being trained. I was surprised: no medical or public health personnel were there, either as students or teachers. The trainees were being instructed on how to deal with a "chembio" event. In simple terms, this entailed evacuating the site, washing down the area and those who had been exposed, providing emergency antidotes if available, and evacuating the victims to a hospital. It was an appropriate program for handling a chemical attack, but it had no relevance for dealing with a biological attack. Such an attack would result in a completely different scenario. Most likely, the organisms would be released as an aerosol—a silent, invisible, odorless cloud. In the case of smallpox, none of those infected would show symptoms for at least a week. There would be nothing to decontaminate. The role of "first responders" would be entirely different. In fact, the true "first responders" would be those in the emergency rooms of hospitals: physicians and nurses would be caring for very sick patients with perplexing symptoms wholly unfamiliar to them. Yet nothing was being done to educate medical staff at hospitals as to the diagnosis and treatment of diseases such as smallpox or anthrax, or to educate public health workers in the investigation and control of an epidemic caused by an entirely unfamiliar disease.

More serious was the fact that most medical and public health professionals did not want to contemplate the possibility of an attack from smallpox or another agent. At that time, the health care community condemned biological and chemical weapons as "morally repugnant," and most declined to participate in research or educational activities that pertained to their potential use. There were only two virology laboratories in the country that were capable of diagnosing the smallpox virus. No smallpox research had been conducted for more than a decade, and the only available smallpox vaccine in the United States had been produced in 1978. There were no production facilities anywhere in the world!

Apathy prevailed. Many leaders in the health care field justified this disinterest by saying that technologically the science of producing and dispersing biological agents was beyond the capability of all but the most sophisticated laboratories. The fact that there had been no known incidents since World War II confirmed for them that there was little cause for concern.

A symposium changes minds

Early in 1997 I left my position as deputy assistant secretary with DHHS to return to Johns Hopkins. I had been commuting four hours a day to Washington for seven years. Enough was enough. About this time, I had several discussions about terrorism and smallpox with Dr. Michael Osterholm, an epidemiologist in the Minnesota Department of Health. Osterholm, a frequent consultant on infectious diseases for federal agencies, was particularly apprehensive about the smallpox threat and interested in bringing public health and medicine into the preparedness planning. Also deeply troubled was Dr. Joshua Lederberg—Nobel laureate, former president of Rockefeller University, and respected consultant to the intelligence and defense communities. He persuaded the *Journal of the American Medical Association* to publish, in September 1996, a special series of articles on the threat of biological weapons, which he edited. The *Journal's* circulation was more than 300,000, but the articles stirred little interest.

A Hopkins colleague, Dr. John Bartlett, then director of the Johns Hopkins Hospital Division of Infectious Diseases and president of the Infectious Diseases Society of America, agreed that the threat was of grave concern, requiring urgent action. He convened a special symposium on the looming menace of bioterrorism at the society's September 1997 annual meeting. The well-known author Richard Preston, Dr. Michael Osterholm, and I presented papers to a standing-room-only meeting of about 3,000. Preston had recently published a widely read book, *The Hot Zone*, which described an outbreak of the highly lethal Ebola virus hemorrhagic disease. It had occurred among laboratory monkeys in Reston, Virginia (a Washington, DC, suburb).

We pointed out to the physicians in attendance that it would be the members of the audience—the infectious disease specialists, emergency room physicians, and intensive care personnel—who actually would be the "first responders" following an aerosol release of smallpox virus. An attack would not be detected until extremely ill patients started showing up in hospitals' emergency rooms. How many physicians were there, we asked, who would be able to diagnose—or would even suspect—a case of smallpox, or anthrax, or plague? How many would know what they should do to treat the victims, or to deal with the epidemic?

An expert "working group"

That symposium was a tipping point. Over the next six months I was invited to give more than fifty additional presentations to professional societies, to medical center staff, and to lay groups. With so few medical and public health personnel who were knowledgeable in this field, I soon had more in the way of correspondence, interviews, and papers to handle than I could manage.

A signal event that followed the symposium was a "table-top exercise" in New York City on a Saturday. It was attended by Mayor Rudolph Giuliani and his staff and was designed to work through the issues following an aerosol release of anthrax. It became evident that the designers of the exercise knew little about public health and were hopelessly misinformed about anthrax. Public Health Service and US military representatives in attendance agreed that an authoritative manual was needed that could serve as a basic text for those in medicine and public health and that would cover the agents of greatest concern. What was not agreed upon was whose job that should be. None saw it as their own agency's responsibility. Moreover, convening a committee that could develop recommendations for a manual would require at least nine to twelve months of administrative preparatory work to meet standard federal advisory committee requirements.

There was, however, another route. The key was to have a private entity, such as a university, convene a group and invite others to attend at their own expense. Thus, in May 1998, I convened at Johns Hopkins the first of several meetings of an expert, informal "Working Group." The group was composed of some twenty-five people—academics as well as health department officials at the federal, state, and local levels.

The working group concluded that five biological agents, plus certain hemorrhagic fever viruses, if used as bioweapons had the potential to cause catastrophic epidemics. There was no question but that there were

other organisms capable of causing major outbreaks, but the ones identified were serious enough to threaten the integrity and functioning of
government. The group advised that priority be given to these agents in
developing educational materials and diagnostic tests and in procuring
countermeasures such as antibiotics and vaccines. These came to be known
as the Class A agents.

Of greatest concern were smallpox and anthrax. Smallpox was
selected because of its high fatality rate, its capacity to spread widely, the
large number of vulnerable people, the lack of available vaccine to curtail
outbreaks—and, not least, the high ranking given to it by Soviet biological
weapons experts. Anthrax was chosen because, even though it does not
spread from person to person and antibiotics were available to treat and
prevent the disease, the organism was easier to obtain, grow, and disperse
as an aerosol. The working group reached consensus decisions about how
to manage the medical and public health aspects resulting from a release
of each of the Class A agents. These were published in the *Journal of the
American Medical Association*. Together, they constituted the only relevant
text that was then available.

Only two of the working group meetings had been held when I received
a call from DHHS, asking the group to recommend materials that should be
put into an emergency stockpile. The recommendations were needed urgently
—President Clinton was requesting additional funds to support special
DHHS programs dealing with bioterrorism. We proposed that the US government procure 40 million doses of smallpox vaccine and enough antibiotics
to treat 2 million people exposed to anthrax. In May the president proposed
that $175 million be included in the DHHS budget to their procurements.

From working group to a center

Throughout the summer of 1998, members of the working group, along
with many medical and public health leaders, encouraged me to establish
an academic center devoted to bioterrorism. Efforts to inject sound science
into initiatives to counter bioterrorism were not getting needed traction. A
center, my colleagues believed, could attract a critical mass of expertise
and activity. Fruitful discussions ensued with the US senator from Maryland, Barbara Mikulski, who earmarked one year of funding for such a
center in the DHHS appropriations bill.

Figure 72. Drs. Tara O'Toole and Thomas Inglesby. O'Toole is director and chief executive officer of the Center for Biosecurity at the University of Pittsburgh Medical Center (UPMC). A specialist in occupational medicine, she was formerly assistant secretary of energy in the Clinton administration. **Inglesby** is chief operating officer and deputy director of the Center for Biosecurity of UPMC. He is an internist and specialist in infectious diseases.

I had already been fortunate in recruiting two of the most creative and determined people with whom I have ever worked. They have continued their leadership roles for more than ten years. The first, Dr. Tara O'Toole (see figure 72), a physician trained in internal medicine and occupational medicine, had recently resigned as assistant secretary of energy. We had become acquainted during my time at DHHS. She was a dynamo who agreed to work part-time. (She had pointedly asked, however, whether my definition of part-time was now sixty hours or eighty hours a week.) Legitimate full-time would have been my preference, but funding was a problem. The only resource I had was a modest personal award from a foundation intended to be used to promote vaccination. The second person recruited was Dr. Thomas Inglesby (see figure 72), an Infectious Disease Fellow at Johns Hopkins Hospital, formerly the William Osler Chief Resident in Medicine. He had attended a lecture on bioterrorism that I had given, recognized the importance of the challenge, and asked to work with

Figure 73. Drs. Monica Schoch-Spana and D. A. Henderson. The author with Schoch-Spana, who is a social anthropologist and senior associate with the Center for Biosecurity of UPMC.

me full-time. He said he would do anything to help. I was concerned that my colleague, Bartlett, would be unhappy about my recruiting one of his most promising infectious disease fellows. However, Bartlett was delighted and actually volunteered to cover Inglesby's salary and travel expenses. O'Toole and Inglesby were soon joined by Dr. Monica Schoch-Spana (see figure 73), a recent Hopkins doctoral graduate in cultural anthropology, and Ms. Molly D'Esopo, who brought an extraordinary capability to manage all things administrative. Even with a five-person center, there were more demands than we could possibly handle but precious little money in our account.

With only one year of guaranteed funding, I had to find resources that would permit us to expand the center and keep it going for at least three years. I believed that our work should be attractive to many foundations. From my deanship years, I was acquainted with a number of foundation officers, but when I discussed our plans and needs, their responses were not encouraging. I had difficulty understanding why. Biological agents— whether in the form of weapons or of newly emerging diseases, such as

AIDS—were increasingly recognized as a major peril; our activities were receiving considerable publicity and praise, and our sponsoring institutions, the Johns Hopkins Schools of Medicine and Public Health, were world renowned. And yet our proposals were all rejected, usually with the explanation that the respective foundation board would be queasy about funding studies in such a controversial area. I turned then to foundations that primarily supported policy centers concerned with issues such as détente and arms control. Again, the proposals were flatly rejected—this time, on the grounds that the foundations had no experience in funding programs in schools of medicine or public health.

Eventually, I turned to Dr. Ralph Gomory, president of the Alfred Sloan Foundation. I had come to know and respect Gomory during meetings of the President's Council on Science and Technology and had sensed that he comprehended the challenge posed by biological agents. I explained what we were hoping to achieve. He asked only a few questions, the last being: "What other foundations are supporting the center?" I had to admit there were none. His unexpected response was "Good. We will support you. Ask for what you need. This is an important problem." In October 2000, we received a generous two-year grant from the Alfred P. Sloan Foundation. Since then, continuing, steadfast support by the foundation, by Gomory personally, and by program director Dr. Paula Olsiewski has facilitated a growing agenda that has contributed substantially to national initiatives in public health preparedness.

THE CENTER FOR CIVILIAN BIODEFENSE STUDIES

Our center focused on the threats posed by biological weapons and also on biological security issues concerning newly emerging diseases such as pandemic influenza, human immunodeficiency virus, and monkeypox. We were, in fact, the only policy center that dealt with biosecurity issues and had a staff that included a broadly knowledgeable medical and public health staff. There were so many complex issues at stake and so few of us that we tried to spark interest at other key academic centers in creating similar organizations. There was little interest.

We had two daunting goals: to develop a broad base of political and professional understanding of the threat of biological agents, and to formulate strategic plans to deal with the various threats. An effective program would require the active involvement of people from many spheres: officials from all levels of government, academic leaders, health care providers, members of the private sector, volunteers, the media, and entire communities. Where to begin?

A national symposium for medical and public health professionals seemed to be a logical point of departure. We decided in October 1998 to sponsor a national meeting to be held in Washington four months later. Given the short lead time, it was a difficult challenge—in retrospect, it was all but foolhardy. Deciding on a program, contacting a potentially interested audience, and finding, on short notice, an auditorium to seat one thousand people in Washington resulted in long days and sleepless nights. Five weeks before the meeting, all of twenty people were registered to attend. Then the registrations began to pick up, a trickle became a deluge, and a week before the meeting, we were fully booked. With renowned speakers, attendees came from forty-six states and ten foreign countries and major media covered the story. It was a significant step forward.

Because so few had expertise in this area, we were repeatedly solicited for briefings by different departments of government, the Congress, and medical centers. Invitations flooded in to participate on advisory boards to the intelligence community, the military, and various professional societies. In November 2000, we convened a second national meeting. It was as successful as the first.

THE "DARK WINTER" EXERCISE

The center needed a more compelling mechanism to convey clearly the staggering impact that biological weapons could have. Until members of the US Congress and others at the highest levels of government fully appreciated the potential for disaster and the need to be prepared, they would not commit the necessary resources. We decided that the best way to illustrate this was to dramatize it. In June 2001, with the collaboration of the Center for Strategic and International Studies (CSIS), the Homeland Security Institute (ANSER), and the Oklahoma Memorial Institute for the Prevention of Terrorism, we developed and ran an exercise called "Dark Winter." The potential destruction of a nuclear explosion was known to everyone—but few could imagine the repercussions from a major epidemic of a deadly infectious disease. What would happen if a smallpox epidemic suddenly appeared, afflicting hundreds and then thousands of people, killing three of every ten despite good medical care—and with no more than 15 million doses of vaccine available in the United States to provide protection?

The concept of a senior-level policy exercise that simulated a covert smallpox attack originated with John Hamre, president of CSIS and formerly deputy secretary of defense. Its authors were Tara O'Toole and Tom Inglesby, with Randy Larsen and Mark De Mier of ANSER. Dark Winter was the first exercise of its kind.

Scenario: *The date of the attack scenario was December 9, 2002. Early reports of people infected with smallpox were being received; most were from Oklahoma, but a few cases had been reported in two other states. Preliminary information suggested that an aerosol attack using smallpox virus probably occurred around December 1 at several shopping malls in Oklahoma.*

The information was being received by participants taking the roles of the US president and his National Security Council in an emergency meeting. The governor of Oklahoma, who happened to be visiting Washington, was asked to join the group. The role of the president was assumed by former Senator Sam Nunn; David Gergen acted as National Security Adviser; James Woolsey played the director of the Central Intelligence Agency; Governor Frank Keating of Oklahoma played himself, the governor of the state where the first attacks occurred. Other roles were taken by individuals who had formerly held high-level government positions. Five senior journalists from major news networks covered the event and participated in a mock press conference afterward.

Emergency vaccination programs were critical to control the developing outbreaks—but the country had only 15 million doses. What, then, was the best way to use this vaccine? Should it be given only to contacts of patients and to essential personnel? Should it be distributed more widely in the three affected states, or should some vaccine be provided to all states? There were other serious questions. Should state borders be closed? What about international borders? Should the National Guard be mobilized? Who should take the lead in communicating information to the public, and what should that information be? Over the next two weeks, the number of cases grew and more states became infected. Television announcements were made and messages came in from intelligence sources and from other countries. The clock was moved ahead to examine the state of problems on December 15 and, finally, on December 22. The number of infected people grew to 16,000, vaccine supplies were depleted, borders were closed, hospitals in some areas had exhausted their isolation facilities. Schools and stores shut down.

Participants and observers alike were clearly taken aback by the range and complexity of problems that would be triggered by a smallpox epidemic. At the core, it was a matter of medicine and public health, but there were profound implications—political, ethical, and legal—and troubling issues of foreign policy and national security. The exercise brought home many unpleasant realities to the participants. It was apparent that the United States was ill prepared to deal with such an event. Concern about the threat of biological agents rose sharply.

Media coverage of Dark Winter was extensive. There were numerous special briefings of senior members of the administration and congressional leaders; six congressional hearings were held. In fact, on September 6, 2001, I was with former Senator Sam Nunn at a special briefing for members of the Senate. Then came September 11.

NEW CHALLENGES—POST–SEPTEMBER 11

The attacks of September 11 reinforced fears about a biological attack. On Sunday afternoon, September 16, I received a telephone call from Secretary Thompson's office asking me to attend a 7:00 p.m. meeting that night in his office. For me, this was an unexpected call. I had never met the secretary and I had no notion of the agenda.

The principal topic was the looming possibility of a biological weapon being deployed. Officials feared that there might be a second terrorist attack, and some considered smallpox virus or anthrax to be the most likely weapons. The new administration had been in office for only eight months. Little thought had been given to the prospect of biological weapons. The meeting lasted until nearly midnight.

Scarcely a month later, the national level of anxiety about biological weapons rose almost to panic level when severe inhalation anthrax cases began to occur. Letters laced with anthrax spores resulted in eleven severe cases of whom five died. The terrorist's identity was not known at the time (and it still is unclear). Additional attacks had to be anticipated.

Some preparations had been undertaken since our working group meetings in 1997. An assistant secretary for DHHS had been recruited— the experienced and very capable Dr. Margaret Hamburg, formerly director of the New York City Health Department. She provided an initial impetus to planning and preparedness before leaving government with the change in presidents. The DHHS budget for bioterrorism preparedness had been raised from virtually nothing to $300 million in 2001. Some staff had been recruited by the CDC, and some funding had been provided to the states and local preparedness programs. Laboratory diagnostic capabilities had been increased, and educational programs were beginning. A national stockpile of medical supplies was being assembled with a special emphasis on preventive measures for anthrax and smallpox. For anthrax, there were enough antibiotics to provide preventive treatment for at least 2 million people for sixty days.

For smallpox, vaccine was critical. There was no effective drug for treatment. Funds had been set aside to buy 40 million doses of vaccine, but there were no manufacturers. In the United States the sole vaccine producer through the 1970s had been Wyeth Laboratories. It had produced the vaccine in the traditional manner—by scarifying the skin of calves, harvesting the pustular material, purifying it, and then freeze-drying it for distribution. After worldwide eradication, Wyeth dismantled its production facilities and had no plans to produce smallpox vaccine again.

The needed vaccine would have to be produced in tissue cell culture, a new and unexplored process for the production of smallpox vaccine. Inevitably, there would be delays in perfecting the production methodology and the necessity for special studies of vaccinees to ensure that the vaccine was both safe and effective. However, there was no option but to move ahead.

We would need a new manufacturer. Beginning in 1999 the heads of several companies had been approached about producing a modern tissue cell culture product. Most helpful had been Dr. Thomas Monath, the principal scientist with a small vaccine development company, Acambis. Monath had formerly worked in US Army and CDC laboratories and knew well the characteristics of vaccinia virus and the mechanics of vaccine development, production, and licensure. He understood the urgency of producing large quantities of vaccine quickly, but the fastest development path he could devise would require at least five years. This assumed that all proceeded optimally—which, in vaccine development, it seldom does.

HOW TO DEAL WITH A SMALLPOX EPIDEMIC

After the September 11 tragedy, we tried to determine what we could do if a smallpox virus attack occurred within the next few weeks or months. Potential scenarios were every bit as grim as the one portrayed only four months before in the Dark Winter exercise. The United States had only 15 million doses of smallpox vaccine. It had been produced in 1978 and kept in deep freezers. On September 17, I made a call to the CDC to check on its status and learned that only 90,000 doses were immediately available. Much of the fluid needed to reconstitute the dried vaccine powder had gone bad and would have to be replaced. The CDC would also have to recheck the vaccine potency of the vaccine lots, many of which hadn't been checked in at least eight years. They were supposed to be checked every three years.

We knew that supplies of vaccine elsewhere in the world were limited. The World Health Organization had surveyed countries in 1998 to determine how much vaccine was available—the total was about 80 million doses. Knowing what I did about likely storage conditions in several of the countries, I assumed that usable vaccine would be much less than half that amount. Moreover, we concluded that there was no possibility of obtaining vaccine from other countries if we needed it. No country would release its own supplies of vaccine if smallpox was again abroad in the world.

The potential damage for a large-scale smallpox epidemic in the United States had never been greater. Routine vaccination had stopped thirty years before, and probably 75 percent of the population was now fully susceptible—only a small proportion would still have some remaining immunity. We reviewed the barriers to industrial production but could see no easy way to shorten the time needed to produce vaccine in less than five years. We pondered the possibility of virus research laboratories producing quantities of vaccine. A number of difficulties ruled out this option—including the logistics of quality control, preserving the vaccine for shipment, and distributing it. Under desperate circumstances there was one possible approach that no one liked. This was to revert to the ancient practice of arm-to-arm vaccination in which material is taken from the pustule on the arm of one vaccinee and used to inoculate another person. A major problem in doing this is that other agents such as hepatitis and HIV could be transmitted at the same time. There was no reasonable answer other than to institute emergency measures for producing a new vaccine.

The vaccine production miracle

No one relished the thought of spending the subsequent five years under the cloud of a possible smallpox attack and knowing that little could be done to deal with it. I turned to the one person I knew who would offer the best advice—Dr. Phillip Russell. By virtue of years of work on viruses and vaccines at the US Army research laboratories in Frederick, Maryland, Russell had acquired an extensive and unique knowledge of the intricacies of vaccine development and production. He also had extraordinary skills as a manager and had risen to the rank of major general. He had retired from the military and, when I contacted him, he was serving as a consultant to a number of international organizations and vaccine manufacturing companies when he wasn't harvesting trout or thinning out the moose population. He agreed to work with the department for a few days to chart a course of action. It was clear that the need was dire and the obstacles were formidable. At the end of a week he volunteered to assume full-time responsibility for the vaccine production challenge. Under Russell's tight supervision, a no-nonsense attitude toward all engaged in the process, and uniquely contrived research initiatives, tens of millions of doses of smallpox vaccine began flowing into the stockpile within eighteen months from the Acambis Company working with Baxter Laboratories.

The pressing question was, how much did we require? The working group had proposed stockpiling 40 million doses with provision being made for the emergency production of larger quantities if needed. When kept in a deep freezer, the vaccine is highly stable and can remain potent for decades. We needed a company that could make available a dedicated facility and was prepared to sustain indefinitely a production capability that would permit it to produce additional vaccine under emergency circumstances. Special studies would be needed to test the new vaccine and to determine the optimum strength for it.

The stockpile target rapidly changed, however, when Secretary Thompson, endeavoring to be reassuring at a press conference, stated that the government would stockpile a dose for every person in America. At first, I looked for a way to set a lower target. I believed that 40 million doses in the stockpile with an emergency manufacturing capacity would be sufficient. However, as I thought about it, I changed my mind. The reality was that if smallpox were released anywhere, the United States as well as all other countries would be at risk. We would probably be one of only three or four countries with enough vaccine to help control the epidemic. The entire world would be at risk of a spreading epidemic, and, in our own interests, if not for humanitarian reasons, we would need to participate in emergency control measures in other countries as well—and that meant distributing vaccine.

Russell was a well-organized, tough, knowledgeable, and demanding leader. Every two weeks, a meeting was held of the principals from the DHHS agencies (Food and Drug Administration, CDC, National Institutes of Health), Defense and Commerce Departments, manufacturing consultants, and others. Reports were shared, targets adjusted, and problems solved. Issues that in the normal flow of a bureaucracy would require weeks to sort out were often settled in minutes.

A competitive contract proposal was drawn up and awarded in record time—sixty days—a process normally requiring a year or more. The Acambis Company joined with Baxter Laboratories to produce the vaccine at a plant in Austria and to ship it to the United States where it was bottled and packaged. By the spring of 2003 we knew that we would have at least 200 million doses by the autumn of that year. A great deal of additional testing would be needed before it could be licensed. Indeed, it was not until 2007 that the Food and Drug Administration finally granted licensure. In the interim, however, we were prepared for emergency use of the vaccine should it be needed.

THE OFFICE OF PUBLIC HEALTH EMERGENCY PREPAREDNESS

After my September 16 meeting with Secretary Thompson, I spent increasing amounts of time commuting to Washington as a consultant. In October, he asked that I assume direction of a new office—at the level of assistant secretary—reporting directly to him. The office was to have overall responsibility for the development of civilian readiness to cope with the health and medical aspects of a biological, chemical, or nuclear attack. It was called the Office of Public Health Emergency Preparedness. Some thought the name too bland and argued for something more dramatic—perhaps "War Office against Bioterrorism" or "Supreme Headquarters for Medical Counterterrorism" or some such. I disagreed. What we needed in all areas were community-wide plans, drawing on the full participation of diverse resources—public health and medical personnel; hospitals; the Red Cross and other voluntary organizations; the police, fire, and emergency rescue technicians; the private sector; schools; the media; and others. It would require community-wide organization and planning of a complexity we have never known in the United States. This is uniquely the domain of "public health." In presentations to Congress and the public, the community-wide cooperative effort was what we stressed and the need to strengthen the public health structure. The name of the office served as an important reminder.

The office had a name, a couple of rooms buried deep within the Hubert Humphrey Office Building in central Washington, and almost no staff. For me to find my way through the existing bureaucratic maze would have been impossible had not Dr. William Raub agreed to be my deputy. Raub was a scientist, a veteran of government service, formerly deputy director of the National Institutes of Health, and a colleague at the White House during the President George H. W. Bush years. He also happened to be the finest civil servant whom I had ever known—knowledgeable, imaginative, fully capable of quietly solving the most intractable of problems, totally unflappable, and loyal. Another important figure was Stewart Simonson, special counsel and long-time friend and confidante of the secretary. He was comparatively young and new to government but a quick learner and determined to allow neither bureaucracy nor red tape to stand in the way of getting a job done. Little would have been achieved without both of them.

With my full-time appointment to the office in Washington, I proposed that O'Toole and Inglesby be named director and deputy director, respectively, of our Hopkins Center. They willingly accepted. Two years later, they changed the name to the Center for Biosecurity, and in 2003 it came under the aegis of the University of Pittsburgh Medical Center, although remaining in Baltimore.

The need for my full-time commitment to the new office became apparent on January 9, 2002, when President Bush signed a supplemental appropriation bill giving DHHS $3 billion for public health emergency preparedness. It was beyond any sum I could imagine. Receiving it suggested to me the image of a water-starved desert traveler hit by a typhoon. The secretary cheerfully turned to me and said, "This is all in your account. Don't release it to anyone unless you are convinced that it will be used wisely and well." This clearly was more meaningful in metaphor than reality. We had argued the case that preparedness and response rested primarily on the strength of local and state community preparedness. Thus, more than a third of the appropriation was specifically assigned for this purpose. Funds also were made available to schools of public health for the development of needed curricula and training. Neither the states nor the schools moved with the alacrity that we expected considering the gravity of the threat. Gradually, however, programs have begun to mature, but there is a lot of work yet to be done. Meanwhile, the smallpox agenda continued to require time and attention as never before.

LET'S VACCINATE EVERYONE!

One of my first acts after receiving the supplemental appropriations was to sign a brief, two-page memo approving an expenditure of almost $500 million for smallpox vaccine. I regret I have no photo of this memo. Committing money on this scale was a once-in-a-lifetime experience.

Meanwhile, an interim emergency supply materialized, much to our surprise. A US pharmaceutical company, not now engaged in vaccine production, informed us that it had discovered it had in storage more than 80 million doses of smallpox vaccine. It had been produced in 1958 by a previous owner of its production facility and stored in large containers at −20°C. When tested, it was found to be fully potent and produced acceptable responses in vaccinees. With newly manufactured vaccine well advanced in the production pipeline, I rested easier.

It was apparent that by late 2003, we would have a significant and growing stockpile of smallpox vaccine. However, I was taken totally by surprise when I found that Vice President Cheney and the White House Homeland Security Council had taken a burning interest in undertaking a nationwide vaccination program. I soon began spending many hours each week in meetings with the council and the vice president's staff reviewing the pros and cons of such a program—over and over again. My own views and those of my staff were definitely at odds with those at the White House.

The basic question was that of balancing the risks of vaccination complications against the likelihood of smallpox virus being released and spreading rapidly. My belief was that the probability of smallpox virus being used as a weapon was small. It would be difficult to obtain the virus and not easy to grow. But in a highly susceptible population and without vaccine to prevent its spread, it could become a devastating epidemic with profound consequences. With adequate supplies of vaccine available, we knew from our experiences during the eradication campaign that smallpox could be readily contained. The disease spreads slowly enough that it can be quickly stopped by isolating patients and vaccinating several hundred to a few thousand of their close contacts.

So, why not vaccinate everybody and erase the smallpox threat altogether? This was the prevalent view in the vice president's office. We disagreed. When smallpox vaccination was routine in the United States, the threat of imported smallpox was real and ever present. The occurrence of some vaccine complications was considered to be an acceptable risk, given the more serious risks posed by an outbreak. Smallpox vaccination causes more serious reactions than any other vaccine we have. In the public arena today there is a general expectation that all vaccinations will be essentially painless and harmless except for the initial needle prick and perhaps a slightly sore arm. Serious side effects are almost unknown. Smallpox vaccination, however, results in a pustule on the arm that grows and develops over ten days. The arm is usually sore and red around the pustule, and there will be a day or two of fever. For about one in one thousand people, the vaccine causes more serious problems, including widespread rash over the body, which may be severe enough to require hospitalization. People with a weakened immune system, such as those with cancer or AIDS, can die from vaccination complications. Before 1972, when vaccination was last required, several hundred people required hospitalization each year for a complication of vaccination, and about one in one million died. Vaccination was accepted as a necessary risk when smallpox with its high death rate threatened, but that threat was not now perceived to be very likely.

The dilemma was, what should we do when we had enough vaccine to

THREE PRICKS OR FIFTEEN

One of the most absurd conflicts with which I was involved was over the use of the bifurcated needle in the United States in 2003. The needle had been invented in 1966 and was used in all countries with WHO-supported smallpox programs. Hundreds of millions of vaccinations had been satisfactorily performed with it.

The problem arose from the manufacturer's original 1966 instructions. The two-pronged needle was intended to replace a standard needle in implanting vaccine virus into the skin. Using the standard needle, it had been the practice to make five pressures into the skin for first-time vaccinees and fifteen for revaccinees. When the bifurcated needle was introduced, the manufacturer noted that there were two prongs, and so only three pressures would be necessary for primary vaccinees.

When we adopted the needle for use in the WHO program we modified the technique. Instead of pressing the vaccine into the skin, we turned the needle at right angles to the skin surface and made multiple punctures. We soon discovered that vaccinators, doing only three punctures, would often make them so gently that the prongs did not break the skin. These vaccinations were unsuccessful. If vaccinators did fifteen punctures rapidly, vaccinations were nearly always successful. Fifteen punctures became the standard. This had been the technique used in all of the recent studies of vaccination and was standard procedure in the military and in vaccinating those at risk in laboratories.

All went well until we were preparing final instruction sheets for use in the ill-fated vaccination program. An FDA staff person discovered that the 1966 instruction sheet called for only three pressures for primary vaccinees and "three" was the number we would have to stipulate in our instruction forms. I reminded her of the fact that hundreds of millions had been vaccinated with the fifteen puncture technique and that thousands had been similarly vaccinated by the CDC, the military, and others, including FDA staff. She insisted that proper documentation be provided. Two million dollars was spent assembling data on appropriate forms. Meanwhile, records were checked with the inventor's company, Wyeth Laboratories, and in FDA files, looking for test results comparing the effectiveness of three and fifteen pressures. There were none.

The outcome? The FDA scientist and attorneys were not comfortable with this revolutionary new technique of fifteen punctures for everyone. The instruction sheets went out, calling for three punctures for primary vaccination. A clear victory for bureaucracy over science!

be able to vaccinate everyone? There were four options: (1) vaccinate no one but have the vaccine ready to be used quickly; (2) vaccinate those at high risk—hospital workers in emergency rooms and intensive care units and first responders; (3) make vaccine available to all who want it; (4) make vaccination compulsory. I favored the first option, as did most health care staff. Vice President Cheney and his aides strongly favored some version of the third option, perhaps even the fourth. They saw smallpox as being highly likely, an almost certain catastrophic threat.

In meeting after meeting, we presented our data about the risks of vaccination and about our capability to stop an epidemic. We prepared explanatory papers, expanded them, and presented them again many times to various White House and Homeland Security groups and individuals. I tried to explain how hard it would be to mount a successful national vaccination program unless there was an outbreak. Reaching even half of the population with an innocuous influenza vaccine was a yearly challenge. I reminded everyone of the negative public response and rejection of vaccination when only a few adverse events occurred after certain childhood vaccines. In this case, if we vaccinated even 50 million people, we could expect fifty deaths due to the vaccine. Our arguments had little effect. Additional, often repetitious meetings were called.

One day I was asked, with two of my staff, to accompany Vice President Cheney and two of his staff on a flight to Atlanta on *Air Force Two*. At the CDC, we met with Dr. Julie Gerberding, the director, and her staff. We spent a full day rehashing facts and figures once more. Neither the vice president nor his staff asked questions, nor did they seem to disagree. The vice president exhibited little more than his characteristic half-smile and said very little. About the only comment that I can recall any of them making was "Hmm."

I reached home about ninety minutes after deplaning at Andrews Air Force Base to be met by my wife, who said: "Your office just called. You should be there early tomorrow because they are planning to make an announcement about vaccinating the entire country." I took an early train and walked into my office expecting to deal with press releases and numerous questions and phone calls. A few staff were already there. They were relaxed and chatting. The White House had just called, they said, and the press conference had been canceled.

How or why the decision was reversed is still somewhat of a mystery.

One source thanked me for having talked with President Bush some months before when I flew to Pittsburgh with him and Secretary Thompson on *Air Force One*. During an hour's informal discussion he had asked a number of questions about biological weapons and preparedness. I told him how we proposed to deal with a smallpox outbreak and how we should be able to stop it within no more than three to four weeks and with perhaps a few hundred thousand vaccinations at most. He nodded to indicate that he understood, but he took no notes. Did he overrule the vice president? It is all but certain that he did.

But this was not the end of it.

A national vaccination program starts and collapses

The issue of smallpox vaccination would not die. Within weeks, it was back on the table—this time with a compromise plan announced by the White House. There would be a program, but it would be set up in three phases. The first phase would offer vaccination to an estimated 450,000 staff who worked in the emergency rooms and intensive care units of hospitals and to public health teams who might investigate cases. These were the people most at risk and who might be exposed to infectious patients before smallpox was diagnosed. The second phase would be to vaccinate traditional "first responders"—police, firefighters, emergency rescue workers, and others at risk. The total was thought to be about 10 million. When these two phases had been completed, further consideration would be given to vaccinating others—the third phase.

The program was an abject failure. Quite inadvertently, our office was partly responsible. Much earlier, we had prepared an eight-page, full-color brochure showing on one side what a normal vaccination reaction looked like and, on the other side, graphic depictions of unpleasant possible complications. It was ideal to be displayed on a bulletin board. We deemed this to be a necessary educational effort since most physicians under age sixty had never seen even a normal vaccination pustule. PHARMA, representing the pharmaceutical companies, covered the cost of printing 500,000 copies. The demand was great, and several million were eventually distributed.

Despite initial interest, the program ground to a halt not long after it began—fewer than 40,000 people had been vaccinated. Medical personnel were not enthusiastic about being vaccinated—particularly after seeing the brochure. No one could say who would pay compensation costs for medical care and loss of wages if a complication occurred. A further contributing factor was that during the Iraq invasion in 2003, it had become known that the Iraqis had no stores of smallpox virus. Many hospital and public health directors took this as an indication that smallpox was no longer a serious threat. Officials declined to participate in the program, many potential recipients refused vaccine, and the initiative was quietly dropped.

SMALLPOX ON THE INTERNATIONAL SCENE

Few countries in the world are prepared to deal with an attack that uses smallpox virus or any other biological weapon. Perhaps a dozen countries have a stockpile of smallpox vaccine large enough to cope with an epidemic in their own country, but none have surplus vaccines to share with other countries, except for the United States. In 2005 I chaired a special committee of the World Health Organization to consider developing an international vaccine stockpile. At that time WHO had fewer than 3 million doses in stock—an amount that was far too small to cope with a spreading outbreak. The committee advised that an international stockpile of 200 million doses be created. Since that meeting, little has been donated. There are now only four smallpox vaccine manufacturers in the world, but none of them can meet the demands of a large emergency. Plans have been approved for an emergency standby production facility in the United States, but this is several years away from being operational.

Atlantic Storm—a reminder that smallpox is not to be forgotten

Concerns about biosecurity have been increasing in many industrialized countries, and some, including the United States, have begun to implement national response plans. Coping with a biological attack in one country is a problem; coping with an attack affecting several countries presents challenges still to be resolved—allocating scarce resources such as vaccines among countries, managing the movement of people across borders, providing coherent reports to the public, and advising people what they should or should not do. The transnational implications of a bioweapons attack had never been explored. Accordingly, our Center for Biosecurity planned a

Figure 74. Dr. Madeleine Albright. The former US secretary of state is shown here in the role of president of the United States in a fictional scenario called Atlantic Storm. *Photograph courtesy of Kaveh Sardari.*

table-top exercise for key political leaders from Europe, Canada, and the United States. It was held in Washington on January 14, 2005. The central theme was the implications of an attack with smallpox virus by terrorists on several countries in Europe and North America. The exercise was called "Atlantic Storm" and was planned with the Johns Hopkins Center for Transatlantic Relations and the Transatlantic Biosecurity Network, with support from the Alfred Sloan Foundation and the Nuclear Threat Initiative.

The setting for the scenario was a ministerial-level meeting of summit leaders from nine countries, the European Commission, and WHO (see figure 74). The time was the present, January 14, 2005. Those who played the roles of heads of state had once held the same or other high-level position in their own countries. Former secretary of state Madeleine Albright took the role of president of the United States; Dr. Gro Brundtland, former director general of WHO and prime minister of Norway, played the part of WHO director general. Others included Bernard Kouchner, soon to become the French foreign minister, and Jan Eliasson, who played the role of prime minister of Sweden and would later serve as president of the United Nations General Assembly. Atlantic Storm was filmed in its entirety by the BBC and included a mock press conference covered by six foreign correspondents.

Scenario: *On the eve of a summit meeting on transatlantic security, reports are received of bioterrorist attacks in Europe. The leaders decided that, before returning home, they would discuss the responses they should take. The exercise began with early reports of smallpox from Istanbul, and subsequently in Frankfurt, Warsaw, Rotterdam, New York, and Los Angeles. The virus had been spread covertly using aerosol sprayers. As further reports came in, it became apparent that there was an international crisis. A number of problems arose, extending well beyond medicine and public health into political, security, and diplomatic areas. The number of smallpox cases increased steadily and began affecting more countries, some of which had no vaccine. None had enough to make it readily available to others. How should the available stocks be used for the maximum benefit, and who could decide? There was no ready communications plan for the countries to exchange up-to-the-minute information or to coordinate action. Some leaders wanted to close borders; others foresaw an economic catastrophe if they sought to do so. WHO was woefully lacking in capacity to respond in any meaningful way, but there was no other international organization that, under the emergency conditions of an epidemic, could provide the needed forum, communication links, and organizational resources.*

The participants recognized that in facing an unfolding epidemic, resources were far short of what they needed, the options were few, the choices were stark, and the potential implications were dire. They recognized the need for the development of national plans and preparation. Equally important, however, was the need for the international community to develop coordinated strategic and operational plans. A greatly strengthened World Health Organization could play a critical role in facilitating a multinational planning process, in building emergency stockpiles, in aiding better communications, and in mobilizing necessary resources. Would countries support such a role and fund it adequately? That is still unknown.

Enormous changes will be needed if the world community is to be prepared to cope with what many feel are inevitable crises engendered by biological weapons attacks or the emergence of rapidly spreading and deadly viral or bacterial infections. Meanwhile, smallpox continues to hover as a dark and ominous cloud as it has throughout the course of human history. It can neither be forgotten nor ignored.

Chapter 11

LESSONS AND LEGACIES OF SMALLPOX ERADICATION

"Familia sic putant omnes quae jam factor, nec de salebris cogitant, ubi via strata.

When a thing has once been done, people think it easy; when the road is made, they forget how rough the way used to be."
—Robert Burton, *The Anatomy of Melancholy*, 1621

The dramatic victory over one of our oldest, most feared, and devastating diseases was a landmark achievement by dedicated, imaginative smallpox eradication staffs from nations across the world. Under the aegis of the World Health Organization, all countries joined together in striving toward a common goal. International field staff never numbered more than 150, but they played key roles in catalyzing national programs. Regularly, they devised new solutions to surmount the many almost intractable obstacles. National staffs numbered in the thousands, but during the climactic end stages, as many as 150,000 were engaged. They likewise exhibited remarkable ingenuity and persistence in tackling new and difficult problems.

The great campaign had begun in 1967 when there were at least 10 million smallpox cases each year. Little more than ten years later, the last case occurred, enabling vaccination to be stopped everywhere. The prevalent reaction around the world was disbelief. How could one be confident that there were no cases anywhere—not in trackless jungles, nor in teeming urban slums, nor among weary refugees fleeing floods or civil war? Tens of thousands of rumors were patiently investigated. Large areas with limited health services were searched village by village and house by house. A generous reward was posted to be given to anyone finding a case. After October 26, 1977, no cases could be found.

As I have described, the challenges were more difficult and the solutions far more complex than one might imagine. There were weeks and months when the outcome hung in a precarious balance. On several occasions, the ultimate success of the program itself had to be doubted. Miraculous effort and unselfish sacrifice retrieved the programs on a number of occasions. Thus, I have been astonished by colleagues who have suggested that smallpox eradication must have been easy—simply a matter of establishing a command post at Geneva's WHO headquarters and supplying vaccine to armies of vaccinators in the field. This type of statement has often been a prelude to the speaker's proposal that another disease should now be targeted for eradication. Indeed, some with more grandiose dreams have argued that a whole new vista for public health has opened—the eradication of many diseases. To me, this is the wrong lesson to be absorbed.

THE SIREN SONG OF ERADICATION

In 1980, less than six months after the declaration of smallpox eradication, a meeting on eradication was convened at the US National Institutes of Health (NIH). The subject was what disease should be eradicated next. I was invited to be the keynote speaker. It was anticipated that I would be an enthusiast for tackling other new and daring projects. I disappointed the organizers. I stated that at that time, I could identify no disease that possessed attributes that would qualify it as a suitable candidate for eradication.

During the course of the smallpox program, our staff did not speculate about the possibility of eradicating another disease. From personal experience, they knew that smallpox eradication had barely succeeded. And yet, the potential for smallpox eradication had been so much greater than for any other disease. A political commitment, in both industrialized and developing countries, was driven by the severity of the disease, but we also had the advantages of an inexpensive, heat-stable vaccine, a simple bifurcated needle to deliver it, an easily diagnosed disease with no animal reservoir or asymptomatic carriers, and the feasibility of an effective surveillance-containment strategy.

The world community has experienced demoralizing failures in past programs to eradicate malaria, yellow fever, yaws, and hookworm disease. The collapse of the malaria eradication program was the most recent. That program had begun in 1955, but by 1967 it was acknowledged to be failing just when smallpox eradication had begun. At its outset, the malaria eradication program was characterized by highly optimistic predictions of success, a significant commitment of resources, and rigid as well as exquisitely detailed plans of operation. When the initiative was finally abandoned, more than 2 billion dollars (1960 dollars) had been expended and the credibility of both WHO and public health had been seriously compromised. Not surprisingly, the subsequent proposal that smallpox eradication be undertaken was viewed with skepticism and doubt.

At the 1980 NIH global eradication meeting, all manner of diseases and conditions were proposed for consideration. These included poliomyelitis, measles, neonatal tetanus, urban rabies, dracunculiasis (Guinea worm disease), tuberculosis, leprosy, malaria, hunger, and traffic accidents, among others. The meeting was more evangelistic than scientific.

Two diseases were eventually chosen—dracunculiasis in 1986 and poliomyelitis in 1988. Both were targeted for eradication by the year 2000. Unfortunately, after more than twenty years in operation, neither program has yet succeeded. It has proved difficult, as we had discovered, to sustain the impetus for a global program in many different nations with varying interests in the outcome and for long periods of time. Donor and staff fatigue as well as civil war in several of the remaining endemic countries now compromise the likely success of both campaigns. In addition, the polio campaign is now confronting serious technical issues that were not anticipated at the time of its launch.

A 2002 report of the US Institute of Medicine on eradication stated, "We must be cautious about embracing the ideology of eradication at the expense of sound, rational judgment." It is sensible advice. At this time, I don't believe we have either the technology or the commitment to pursue another eradication goal. More useful and contributory would be to build and sustain effective control programs that are adapted to the social and public health needs of each country.

THE LEGACY OF THE SMALLPOX ERADICATION PROGRAM

I believe that the important, longer-term contribution of smallpox eradication to world health was its demonstration of how much could be accomplished with how little in the control of infectious diseases through community-wide vaccination programs. The success of smallpox eradication led to a surge in such programs and greatly expanded research efforts targeting the development of new vaccines. The WHO Expanded Program on Immunization (EPI) was a major driving force.

Observations early in the eradication program had set the EPI in motion. In 1968–1969, during my early visits to infectious disease hospitals, I was surprised to find whole wards given over to the treatment of common preventable childhood infectious diseases. Yet, inexpensive vaccines for a number of these were widely available in industrialized countries. Most developing countries used few or no vaccines other than smallpox vaccine. Where other vaccines were employed, coverage was seldom greater than 10 to 15 percent. At the same time, we were discovering that with good management, it was possible to vaccinate successfully and inexpensively large numbers of children against smallpox using small, mobile teams. In well-planned African programs, a team of four vaccinators could average 2,000 smallpox vaccinations per day—400,000 vaccinations per year. With the help and cooperation of village leaders, a coverage of 80 percent or more could be expected. Supervisors traveling with each team ensured that vaccinations were properly performed and that the vaccine was appropriately handled so as to retain potency.

It seemed to me that a well-organized smallpox eradication program could serve as a basic structure for the delivery of other vaccines and perhaps other interventions intended to reach citizens throughout a country. For the smallpox program, we needed systems for vaccine procurement, storage, and distribution; methods for quality control of vaccines and vaccinator performance; staff trained in obtaining villagers' cooperation; simple record systems for recording vaccinations; and a national disease surveillance system to help evaluate progress and to guide strategy and tactics. To expand the smallpox vaccination program and to incorporate other vaccines would require additional funds as well as new methods for the storage, transport, and distribution of less-heat-stable vaccines. The feasibility of giving measles and smallpox vaccines simultaneously was already being demonstrated in the West Africa program. Why not extend this approach to include other vaccines?

An opportunity to introduce the concept of an expanded vaccination initiative came in 1970, when Dr. Charles Cockburn, WHO director of viral diseases, and I were asked to arrange an international conference on the use of vaccines against infectious diseases. It was convened at the Pan American Health Organization (PAHO) in Washington. As part of the conference, a committee of participants was formed to develop a recommended immunization program for the developing countries. Dr. Julie Sulianti Saroso, director-general of the Indonesian Ministry of Health, was chairman, and I served as secretary. The committee recommended the universal use of a number of vaccines in addition to smallpox—vaccines protective against diphtheria, whooping cough, tetanus, measles, and tuberculosis, as well as polio and yellow fever vaccine in selected problem areas. The economies and advantages of using mobile teams were emphasized. The report named this initiative the Expanded Program on Immunization (EPI).

Endorsement and support by the World Health Assembly was needed, but little progress was made until 1973, when Dr. Halfdan Mahler became director-general. A proposal was presented and approved by the 1974 World Health Assembly. I was asked to assume direction of the new program in addition to smallpox eradication. This request came at the time when smallpox programs in India, Bangladesh, and Ethiopia were reaching crisis stage. I was then spending close to 70 percent of my time traveling and, like most staff, I was already working seven days a week. I feared that assuming responsibility for two programs could well spell the demise of both. Regretfully, I declined.

Early efforts to begin the program floundered. Too few staff and resources were provided. Moreover, it was not an auspicious time to expand routine smallpox vaccination programs to include other vaccines. Programs of routine smallpox vaccination were scaling down. The Americas and Indonesia had already been certified as having eradicated smallpox, and personnel there and in West Africa were focused primarily on confirming that smallpox transmission had been interrupted. The few remaining endemic countries were fully preoccupied in stopping the last outbreaks. Few voluntary contributions for EPI were forthcoming.

In meetings with Mahler, I expressed my concern and argued for a strengthened WHO office. He agreed, and so, in the spring of 1976, a more ambitious role for WHO was laid out. It included a professional EPI head-quarters staff of no fewer than twelve people to work with donor countries, to develop training programs, to organize a research agenda, to chart the logistics for transport of temperature-sensitive vaccines, to travel to countries to ensure the development of quantitatively targeted plans, and to put together surveillance programs. I emphasized the need for a broadly integrated program throughout WHO with dedicated regional office staff working with those at headquarters and with national program officers in each country. It seemed logical that, with the success of smallpox eradication, WHO should continue its lead in advocacy and execution of health programs rather than acting as a passive observer.

AN EXPANDED PROGRAM ON IMMUNIZATION BEGINS

Fortuitously, a new assistant director-general for communicable diseases, Dr. Ivan Ladnyi, arrived on the scene. Until 1972 Ladnyi had been our very productive regional smallpox eradication adviser for countries in eastern and southern Africa. He had returned to a senior position in the Soviet health ministry but was recruited by Mahler four years later to be a WHO assistant director-general overseeing WHO's infectious disease programs. Recruitment in 1977 for a new director of EPI brought Dr. Ralph Henderson (no relation to the author), then at the Centers for Disease Control (CDC) in Atlanta. He had previously been deputy director of the CDC West Africa smallpox eradication program and later the CDC's national director of venereal disease control activities.

A massive task lay ahead as the staff sought to work out methodologies for conducting programs initially involving DTP (diphtheria, tetanus, pertussis), measles, polio, and BCG (tuberculosis) vaccines. The history of the evolution of the EPI program is a fascinating and tumultuous one beyond the scope of this book. Suffice it to say, substantial support for the program grew rapidly beginning in the early 1980s as the United Nations Children's Fund (UNICEF) decided that immunization programs should be a priority effort, and Rotary International began to raise large sums of money for the purchase of oral polio vaccine. The aim was to achieve, throughout the developing world, 80 percent vaccination coverage by 1990. The goal was met in most countries.

A paradigm for EPI—the program in Latin America

The most dramatic progress in EPI has been in Latin America. PAHO and the countries of the Americas more than met my most hopeful expectations. As chairman of its technical advisory group from 1985 to 2003, I had the opportunity to closely follow its progress.

The PAHO EPI exemplifies the best of what can be accomplished by an international organization and its participating member countries when there is competent executive leadership. Such was provided, in this instance, by successive regional directors (Dr. Carlyle Macedo from Mexico followed by Sir George Alleyne from Barbados); inspired technical leadership by PAHO regional immunization advisers (Dr. Ciro de Quadros, from Brazil, formerly director of smallpox field operations in Ethiopia, and Dr. Jon Andrus, on detail from the CDC); and support from international staff and national leadership throughout the Americas.

In 1977, a PAHO Directing Council Resolution established EPI and appointed de Quadros its director (see figure 75). During the first three years, all countries appointed national EPI managers who participated in special training programs. A new monthly publication, the *EPI Newsletter*, began distribution and has continued since.

A pioneering initiative was the creation of a revolving fund for the cooperative purchase of vaccines for public health programs throughout the Americas. Countries submitted annual estimates of their needs for vaccines, which PAHO then put out for competitive bids. Shipments were made at requested intervals, and the country, within sixty days after vaccine delivery, reimbursed the revolving fund for the cost. Purchases increased from $2.3 million by nineteen countries in 1979 to more than $300 million by thirty-eight countries in 2008. Cost savings achieved by large-scale purchase and planning were between 70 and 85 percent. Manufacturers were better able to plan production schedules, set up emergency shipments, and ensure the quality of vaccine. Regional confidence and trust grew steadily.

There was motivation for countries to participate in the fund, but to do so, they had to meet certain requirements: (1) a realistic and comprehensive annual national action plan; (2) sufficient infrastructure for vaccine storage and distribution; and (3) a designated national immunization program manager.

Figure 75. Ciro de Quadros with the Author and Dr. Jesus Kumate
at a meeting of the technical advisory group of PAHO's
Expanded Program on Immunization (1990). **De Quadros** (*right*)
was director of the PAHO program; **Kumate** (*center*) was minister of health of Mexico. *Photograph courtesy of C. Quadros.*

Vaccine coverage for the recommended basic group of vaccines is now
approaching 95 percent. The programs are supported by twenty PAHO
epidemiologists working at national, regional, and subregional levels.
More than 90 percent of all vaccine now used in the region's public health
programs is purchased with national government funds. International contributions for vehicles and other associated costs have been received from
many different donors, some contributed through PAHO, and many
received directly by the countries. To ensure coordination, PAHO established a Regional Inter-Agency Committee, which included UNICEF,
Rotary International, the Inter-American Development Bank, US Agency
for International Development, and other bilateral donor agencies. Meetings with national authorities to work out national action plans and possible contributors complement the regional group efforts.

A new approach to large-scale vaccination was tested by Brazil beginning in 1980—a National Immunization Day—designed to vaccinate all
children under five years of age with polio vaccine. Volunteer groups

joined health department staff in a nationwide effort that succeeded in vaccinating a number equivalent to more than 80 percent of children in the target age range on a single day. The numbers of polio cases plummeted, and, as experience grew, other vaccines were added. Traditionalists at WHO's Geneva headquarters were severely critical of this approach, but, as Brazil had discovered, when vaccination was left to health centers and the private sector, it was difficult to attain a coverage higher than about 50 percent. Many thought that national immunization days every year could not be sustained over time. However, a Brazilian worker said to me in puzzlement: "We celebrate Carnaval every year. Why can't we have immunization days every year?"

The director of PAHO in 1985 appointed a Technical Advisory Group (TAG) to offer advice about introducing other vaccines. Year by year, other vaccines were added to the immunization schedule. The number rose from six in 1985 to fourteen in 2008. In many countries, other initiatives were added—distribution of vitamin A (for blindness prevention) and insecticide-treated bed nets (for malaria control).

The TAG also examined the feasibility of stopping polio transmission throughout the Western Hemisphere and agreed to support this initiative. A target date of 1990 was agreed upon, and a methodology for executing the program was worked out by de Quadros and national staff through practical field experience. The last endemic case in the Americas occurred in August 1991. Meanwhile, in 1988, the PAHO methodology for polio eradication was adopted by WHO and introduced for use throughout the world.

The polio program required the establishment of surveillance reporting systems in every country to detect possible polio cases. When the program began in Latin America, there were only some 500 sites—hospitals and health centers—reporting numbers of cases monthly. Prompt reports of all cases of possible poliomyelitis were needed, and specimens for virus isolations had to be obtained from each case. The number of reporting sites grew to more than 40,000 by 2009. They reported weekly rather than monthly, and other diseases also being reported increased steadily. A regional network of accredited national laboratories was created.

As polio vanished from the Western Hemisphere, the ministers of health proposed in 1994 that they undertake to stop measles transmission. As of 1996 more than 240,000 cases were being reported annually, resulting in a great many hospitalizations and deaths. By 2002 endemic transmission had been stopped—the only cases that were occurring resulted from introductions by visitors from Europe and Asia.

Rubella was next on the list. Rubella virus infections of pregnant mothers were responsible for at least 20,000 babies being born annually with defects due to infection while they were still in the uterus. These included permanently disabling deafness and heart disease. With an extensive surveillance system now reporting both measles and rubella and a network of 150 laboratories able to test possible cases, it seemed logical in 2003 to begin eradication of rubella in the Americas. A target date of 2010 was established and is expected to be met.

A miraculous transformation has taken place in the Americas—with smallpox, measles, and poliomyelitis banished from the continent and with other vaccine-preventable diseases decreasing to an all-time low incidence. Systems for tracking other infectious agents are steadily expanding.

During the thirty years since EPI began in the Americas, the mortality rate among children under five years of age has decreased by more than 60 percent. Equally significant is the increasing political support and confidence of all nations across the Americas; broader avenues for cooperation and understanding have been established.

NEW HORIZONS IN PUBLIC HEALTH

Smallpox eradication was a catalytic force in the now rapidly expanding range of community-based initiatives in infectious disease control and prevention throughout the developing world. At the time the smallpox eradication program began, the need to provide for treatment of the sick and injured dominated the agendas of most health ministries. Hospitals and health centers were needed, but they were costly and required more skilled health personnel than most countries could sustain. Infectious diseases were a major problem, but in the 1970s, smallpox eradication was the sole coordinated global prevention program. It was a surprise to health ministries that a well-organized smallpox program required no more than a few rooms for an office, a comparatively small staff with transport, vaccine, and limited equipment. The programs proved to be politically popular—a tangible indication to people throughout a country of its government's concern. A visible benefit was the closing of smallpox wards and hospitals. Increasingly, assistance agencies and countries came to recognize that population-wide public health programs are a productive investment and well within the capabilities of even the poorest developing countries. In a 1993 World Bank study of costs and benefits of all possible health interventions, immunization was acknowledged to be the single most cost-beneficial of all interventions whether curative or preventive.

The past two decades have witnessed the rapid development of community-based preventive services. The extension of services into villages and neighborhoods has proved to be far more effective than waiting for residents to find their way to inconvenient, over-taxed health centers. International donor support has expanded beyond expectations, and programs to counter HIV/AIDS, malaria, tuberculosis, leprosy, some respiratory and diarrheal diseases, and an array of parasitic diseases are growing rapidly. Vaccines to prevent cancer of the liver and cervix are in the vanguard of many more such vaccines, yet to be defined. Meanwhile creative research programs dealing with the key problems of developing countries have expanded greatly; surveillance programs to measure progress and guide strategies are, at last, receiving deserved attention. A sea change is taking place.

A little-recognized but critical factor in catalyzing such changes has been the participation of innovative and dedicated young professionals, both nationals and internationals. They questioned traditional boundaries perceived as limiting what could and could not be done. They had the courage to take on apparently insoluble problems and smothering, entrenched bureaucracies; they willingly worked long weeks without a break, enduring rigorous living conditions. They worked with a global vision. With research now defining bold, unimagined horizons, new generations of public health staff inherit a world of extraordinary challenge and opportunity.

ACKNOWLEDGMENTS

This book was written in 2008–2009, during which time I was serving as a Distinguished Scholar at the University of Pittsburgh's Center for Biosecurity, located in Baltimore. Special thanks are due to Tara O'Toole and Tom Inglesby, directors of the center, who strongly encouraged me to write this book and who were most generous in granting me time to do so. I am grateful to the Alfred P. Sloan Foundation for a generous grant to support this effort and to the Johns Hopkins Bloomberg School of Public Health for making available residual funds derived from the predecessor center I had founded, the Center for Civilian Biodefense Studies.

Over the past eighteen years, I have been encouraged repeatedly by a number of people to provide this personal account of the smallpox saga—especially by Johns Hopkins faculty, medical students, and preventive medicine and medical residents. Three of the more persistent individuals who deserve a special word of thanks are Richard Danzig, Randy Larsen, and Stewart Simonson. To all, I apologize for having taken fully eighteen years since this effort was first seriously proposed, but the unexpectedly persistent smallpox saga bears much of the blame for the delay.

I am indebted to those who undertook to read the manuscript in full and to offer constructive comments. These include several who played a central role in the smallpox eradication program: Frank Fenner, John Copland, Stanley Foster, and John Wickett. Others who were not part of the program offered helpful, candid views in the role of interested readers: Gert Brieger, Stewart Simonson, and Parker Small. Others who also deserve thanks for reading parts of the manuscript: Jon Andrus, Isao Arita, Ciro de Quadros, Nicole Grasset, Leigh Henderson, Zdeno Jezek, Leo Morris, John Neff, and Mark Prendergast. World Health Organization staff were of inestimable help in identifying original copies of a great many illustrations that had appeared originally in the 1988 historical archive we had written, *Smallpox and Its Eradication.*

Assistance in rewriting was provided by Janet Worthington; Matthew Packard served as a technical assistant. Other members of the center generously offered helpful assistance and historical memories: Monica Schoch-Spana, Molly D'Esopo, Price Tyson, Bruce Campbell, and Tanna Liggens.

Graphic artist Megan Van Wagoner was responsible for preparing all of the maps and graphics.

It is doubtful, however, that the book would have been written without the considerable help of my wife, Nana, who served as critic, copy editor, and morale booster.

Special thanks are due to the executive editor of Prometheus Books, Linda Greenspan Regan, and a Prometheus staff that willingly went out of its way time and again to provide help and assistance. Lastly, a word of thanks to Wendy Keller, my agent.

SOURCES

FOREWORD AND PREFACE

Dubos, Rene. *Man Adapting.* New Haven, CT: Yale University Press, 1965.

Fenn, E. *Pox Americana.* New York: Hill and Wang, 2001.

Fenner, F., D. A. Henderson, I. Arita, Z. Jezek, and I. D. Ladnyi. *Smallpox and Its Eradication.* Geneva: World Health Organization, 1988.

Hopkins, Donald. *The Greatest Killer: Smallpox in History.* Chicago: University of Chicago Press, 2002.

Ogden, Horace G. *CDC and the Smallpox Crusade.* Washington, DC: US Government Printing Office, 1987.

Preston, Richard. *The Demon in the Freezer: A True Story.* New York: Random House, 2002.

Roueche, Berton. "A Man from Mexico." In *Eleven Blue Men.* New York: Berkeley Publishing Corp., 1955, pp. 73–88.

CHAPTER 1. THE DISEASE, THE VIRUS, AND ITS HISTORY

Baldwin, Peter. *Contagion and the State in Europe 1830–1930.* Cambridge: Cambridge University Press, 1999, pp. 244–354.

Crosby, A. W., Jr. "Virgin Soil Epidemics as a Factor in the Aboriginal Depopulation in America." *William and Mary Quarterly* (1976): 289–99.

Dixon, C. W. *Smallpox.* London: J. & A. Churchill, 1962.

Fenn, E. *Pox Americana.* New York: Hill and Wang, 2001.

Fenner, F., D. A. Henderson, I. Arita, Z. Jezek, and I. D. Ladnyi. *Smallpox and Its Eradication.* Geneva: World Health Organization, 1988, pp. 1–363.

Glynn, Ian, and Jennifer Glynn. *The Life and Death of Smallpox.* New York: Cambridge University Press, 2004, pp. 1–189.

Hopkins, Donald. *Princes and Peasants.* Chicago: University of Chicago Press, 1983.

Koplow, David. *Smallpox: The Fight to Eradicate a Global Scourge.* Berkeley, Los Angeles, and London: University of California Press, 2003, pp. 9–57.

MacAulay, Lord B. *The History of England.* Edited by C. H. Firth. London: Macmillan and Company, 1914.

McNeill, William H. *Plagues and Peoples.* New York: Anchor Press/Doubleday, 1976, pp. 105–45, 249–56.

Miller, Genevieve. *The Adoption of Inoculation for Smallpox in England and France.* Great Britain, India, and Pakistan: Oxford University Press, 1957.

Rao, A. R. *Smallpox.* Bombay: Kothari Book Depot, 1972.

Shurkin, Joel. *The Invisible Fire.* New York: G. P. Putnam's Sons, 1979, pp. 15–217.

Tucker, Jonathan. *Scourge: The Once and Future Threat of Smallpox.* New York: Atlantic Monthly Press, 2001, pp. 1–38.

CHAPTER 2. THE WORLD DECIDES TO ERADICATE SMALLPOX

Dixon, C. W. *Smallpox.* London: J. & A. Churchill, 1962.

Fenner, F., D. A. Henderson, I. Arita, Z. Jezek, and I. D. Ladnyi. *Smallpox and Its Eradication.* Geneva: World Health Organization, 1988, pp. 315–419.

Glynn, Ian, and Jennifer Glynn. *The Life and Death of Smallpox.* New York: Cambridge University Press, 2004, pp. 190–99.

Ogden, Horace G. *CDC and the Smallpox Crusade.* Washington, DC: US Government Printing Office, 1987, pp. 7–16.

Roueche, Berton. "A Man from Mexico." In *Eleven Blue Men.* New York: Berkeley Publishing Corp., 1955, pp. 73–88.

Shurkin, Joel. *The Invisible Fire.* New York: G. P. Putnam's Sons, 1979, pp. 217–74.

Tucker, Jonathan, *Scourge: The Once and Future Threat of Smallpox.* New York: Atlantic Monthly Press, 2001, pp. 39–60.

CHAPTER 3. CREATING A GLOBAL PROGRAM

Dixon, C. W. *Smallpox*. London: J. & A. Churchill, 1962.

Fenner, F., D. A. Henderson, I. Arita, Z. Jezek, and I. D. Ladnyi. *Smallpox and Its Eradication*. Geneva: World Health Organization, 1988, pp. 422–590.

Ogden, Horace G. *CDC and the Smallpox Crusade*. Washington, DC: US Government Printing Office, 1987, pp. 13–37.

Tucker, Jonathan. *Scourge: The Once and Future Threat of Smallpox*. New York: Atlantic Monthly Press, 2001, pp. 61–76.

CHAPTER 4. WHERE TO BEGIN? A TALE OF TWO COUNTRIES— BRAZIL AND INDONESIA

Fenner, F., D. A. Henderson, I. Arita, Z. Jezek, and I. D. Ladnyi. *Smallpox and Its Eradication*. Geneva: World Health Organization, 1988, pp. 593–657.

Glynn, Ian, and Jennifer Glynn. *The Life and Death of Smallpox*. New York: Cambridge University Press, 2004, pp. 200–27.

Hochman, Gilberto. "Priority, Invisibility and Eradication: The History of Smallpox and the Brazilian Public Health Agenda." In *Medical History*. http://www.pubmedcentral.gov/tocrender.fcgi?action=archive&journal=228 (to be published in April 2009).

Ogden, Horace G. *CDC and the Smallpox Crusade*. Washington, DC: US Government Printing Office, 1987, pp. 92–104.

Shurkin, Joel. *The Invisible Fire*. New York: G. P. Putnam's Sons, 1979, pp. 274–93.

Tucker, Jonathan. *Scourge: The Once and Future Threat of Smallpox*. New York: Atlantic Monthly Press, 2001, pp. 61–89.

CHAPTER 5. AFRICA—A FORMIDABLE AND COMPLICATED CHALLENGE

Fenner, F., D. A. Henderson, I. Arita, Z. Jezek, and I. D. Ladnyi. *Smallpox and Its Eradication*. Geneva: World Health Organization, 1988, pp. 849–996.

Glynn, Ian, and Jennifer Glynn. *The Life and Death of Smallpox*. New York: Cambridge University Press, 2004, pp. 200–27.

Imperato, P. J. *A Wind in Africa: A Story of Modern Medicine in Mali*. St. Louis, MO: Warren Green, 1975.

Ogden, Horace G. *CDC and the Smallpox Crusade.* Washington, DC: US Government Printing Office, 1987, pp. 37–70.

Shurkin, Joel. *The Invisible Fire.* New York: G. P. Putnam's Sons, 1979, pp. 274–93.

Tucker, Jonathan. *Scourge: The Once and Future Threat of Smallpox.* New York: Atlantic Monthly Press, 2001, pp. 61–118.

CHAPTER 6. INDIA AND NEPAL—A NATURAL HOME OF ENDEMIC SMALLPOX

Basu, R. N., Z. Jezek, and N. A. Ward. *The Eradication of Smallpox from India.* New Delhi: World Health Organization, 1979.

Bhattacharya, Sanjoy. *Expunging Variola: The Control and Eradication of Smallpox in India 1947–1977.* New Delhi: Orient Longman, 2006.

———. *Fractured States: Smallpox, Public Health and Vaccination Policy in British India, 1800–1947.* New Delhi: Orient Longman, 2005.

Brilliant, L. *The Management of Smallpox Eradication in India.* Ann Arbor: University of Michigan Press, 1985.

Fenner, F., D. A. Henderson, I. Arita, Z. Jezek, and I. D. Ladnyi. *Smallpox and Its Eradication.* Geneva: World Health Organization, 1988, pp. 711–806.

Glynn, Ian, and Jennifer Glynn. *The Life and Death of Smallpox.* New York: Cambridge University Press, 2004, pp. 200–27.

Shurkin, Joel. *The Invisible Fire.* New York: G. P. Putnam's Sons, 1979, pp. 300–47.

Tucker, Jonathan. *Scourge: The Once and Future Threat of Smallpox.* New York: Atlantic Monthly Press, 2001, pp. 61–118.

CHAPTER 7. AFGHANISTAN, PAKISTAN, AND BANGLADESH— THE LAST STRONGHOLD OF *VARIOLA MAJOR*

Fenner, F., D. A. Henderson, I. Arita, Z. Jezek, and I. D. Ladnyi. *Smallpox and Its Eradication.* Geneva: World Health Organization, 1988, pp. 659–710, 807–48.

Glynn, Ian, and Jennifer Glynn. *The Life and Death of Smallpox.* New York: Cambridge University Press, 2004, pp. 200–27.

Joarder, A. K., D. Tarantola, and J. Tulloch. *The Eradication of Smallpox from Bangladesh.* New Delhi: World Health Organization, 1980.

Shurkin, Joel. *The Invisible Fire.* New York: G. P. Putnam's Sons, 1979, pp. 274–93, 349–70.

Tucker, Jonathan. *Scourge: The Once and Future Threat of Smallpox.* New York: Atlantic Monthly Press, 2001, pp. 61–118.

CHAPTER 8. ETHIOPIA AND SOMALIA— THE LAST COUNTRIES WITH SMALLPOX

Fenner, F., D. A. Henderson, I. Arita, Z. Jezek, and I. D. Ladnyi. *Smallpox and Its Eradication.* Geneva: World Health Organization, 1988, pp. 997–1068.

Glynn, Ian, and Jennifer Glynn. *The Life and Death of Smallpox.* New York: Cambridge University Press, 2004, pp. 219–23.

Shurkin, Joel. *The Invisible Fire.* New York: G. P. Putnam's Sons, 1979, pp. 371–404.

Skelton, James W., Jr. *Volunteering in Ethiopia: A Peace Corps Odyssey.* Denver: Beaumont Books, 1991.

Tucker, Jonathan. *Scourge: The Once and Future Threat of Smallpox.* New York: Atlantic Monthly Press, 2001, pp. 61–118.

CHAPTER 9. SMALLPOX—POST-ERADICATION

Fenn, E. *Pox Americana.* New York: Hill and Wang, 2001.

Fenner, F., D. A. Henderson, I. Arita, Z. Jezek, and I. D. Ladnyi. *Smallpox and Its Eradication.* Geneva: World Health Organization, 1988, pp. 1103–1286.

Glynn, Ian, and Jennifer Glynn. *The Life and Death of Smallpox.* New York: Cambridge University Press, 2004, pp. 228–45.

Jezek, Z., and F. Fenner. *Human Monkeypox.* Basel, Switzerland: Karger, 1988.

Koplow, David. *Smallpox: The Fight to Eradicate a Global Scourge.* Berkeley, Los Angeles, and London: University of California Press, 2003, pp. 137–227.

Tucker, Jonathan. *Scourge: The Once and Future Threat of Smallpox.* New York: Atlantic Monthly Press, 2001, pp. 119–65.

CHAPTER 10. SMALLPOX AS A BIOLOGICAL WEAPON

Alibek, Ken, and Stephen Handelman. *Biohazard: The Chilling True Story of the Largest Covert Biological Weapons Program in the World—Told from Inside by the Man Who Ran It.* New York: Random House, 1999.

Cole, Leonard A. *The Anthrax Letters.* Washington, DC: Joseph Henry Press, 2003.

Henderson, D. A., Thomas V. Inglesby, and Tara O'Toole, eds. *Bioterrorism: Guide-*

lines for Medical and Public Health Management. Chicago: JAMA and Archives Journals, 2002.

Koplow, David. *Smallpox: The Fight to Eradicate a Global Scourge.* Berkeley, Los Angeles, and London: University of California Press, 2003, pp. 59–103, 137–227.

Preston, Richard. *The Demon in the Freezer: A True Story.* New York: Random House, 2002.

CHAPTER 11. LESSONS AND LEGACIES OF SMALLPOX ERADICATION

International Conference on the Application of Vaccines against Viral, Rickettsial, and Bacterial Diseases of Man. Scientific Publication No. 226. Washington, DC: Pan American Health Organization, 1971, pp. 457–650.

Jamison, D. T., W. H. Mosley, A. R. Meacham, and J. L. Bobadilla, eds. *Disease Control Priorities in Developing Countries.* New York: Oxford University Press, 1993.

Muraskin, William A. *The Politics of International Health: The Children's Vaccine Initiative and the Struggle to Develop Vaccines for the Third World.* Albany: State University of New York Press, 1998.

All population data are from the United Nations Population Division: Population Division of the Department of Economic and Social Affairs of the United Nations Secretariat, "World Population Prospects: The 2006 Revision and World Urbanization Prospects: The 2005 Revision," United Nations. http://esa.un.org/unpp.

INDEX